The Sciences of Roger Bacon's Opus Maius

Through an examination of the *Opus maius*, Roger Bacon's great scientific work of 1267, this book offers insight into Bacon's understanding of the role of science in society and the study and practice of science in the later Middle Ages.

Written at the request of Pope Clement IV (d. 1268), the *Opus maius* contains English Franciscan and polymath Roger Bacon's plans for educational and ecclesiastical reform through the study of arts and sciences, which Bacon saw as having been neglected in the curricula of the major European universities. In writing the *Opus maius*, Bacon wanted to demonstrate that not only were the sciences a solid foundation for learning theology, but they could also improve society and help mankind attain salvation. This volume contains nine essays on the sciences Bacon thought were most important in this scheme, including optics, mathematics, astronomy, alchemy, medicine, and experimental science. Each provides a focused examination of the individual sciences, demonstrating how Bacon understood and practised them, as well as how he envisioned they would fit together into his larger programme of scientific reform.

This thought-provoking collection will be a valuable resource to historians of science, medicine, and philosophy, as well as those in medieval and premodern studies.

Meagan S. Allen is a historian of science, specializing in the medical alchemy of the later Middle Ages. Her research interests lie at the intersection of alchemy, pharmacology, and theology, especially in the writings of the thirteenth-century polymath Roger Bacon. She is the author of *Roger Bacon and the Incorruptible Human: Alchemy, Pharmacology, and the Desire to Prolong Life* (2023), which examines Bacon's desire to use alchemy to create a medicine that would extend human life by centuries. Allen is currently Visiting Assistant Professor in the Department of the History of Science and Technology at Johns Hopkins University, USA.

GPHNP

GLOBAL PERSPECTIVES ON THE HISTORY OF NATURAL PHILOSOPHY

Fostering a new approach to the study of the history of natural philosophy this series aims to expand the discussion on natural philosophy cross-culturally and comparatively by focusing on the philosophical reasoning about nature developed particularly, but not exclusively, in three main cultural settings: Europe, the Middle East, and China. One of the main focal points of *Global Perspectives on the History of Natural Philosophy* is the interplay between philosophical and scientific concepts, stances, and problems arising from the premodern consideration of nature, broadly considered. Accordingly, the series provides a cutting-edge framework in which natural philosophy can be considered from new philosophically meaningful angles. Acknowledging the historical interweaving of philosophy and science of nature, the series publishes monographs and edited volumes dealing with the history of natural philosophy from three methodological perspectives: philosophical analysis, historical reconstruction, and comparative studies. Submitted manuscripts may either examine authors and issues from a specific philosophical tradition or engage comparatively with patterns and problems shared by different cultural settings.

Aristotle and New Spain
Virginia Aspe Armella

Ten Follies: A Journey Around the World in Ten Forms of Madness
Marco Innamorati

Colonial Scholasticism: Interpreting Nature, Society, and Philosophy in Spanish America
Edited by Abel Aravena Zamora and Nicola Polloni

The Sciences of Roger Bacon's Opus Maius
Edited by Meagan S. Allen

The Sciences of Roger Bacon's Opus Maius

Edited by Meagan S. Allen

Routledge
Taylor & Francis Group
LONDON AND NEW YORK

First published 2026
by Routledge
4 Park Square, Milton Park, Abingdon, Oxon OX14 4RN

and by Routledge
605 Third Avenue, New York, NY 10158

Routledge is an imprint of the Taylor & Francis Group, an informa business

British Library Cataloguing-in-Publication Data
A catalogue record for this book is available from the British Library

ISBN: 978-1-032-52239-5 (hbk)
ISBN: 978-1-032-52240-1 (pbk)
ISBN: 978-1-003-40570-2 (ebk)

DOI: 10.4324/9781003405702

Typeset in Sabon
by Deanta Global Publishing Services, Chennai, India

Contents

Contributors *vii*

The *Opus maius*: A Plea for Reform 1
MEAGAN S. ALLEN

1 Roger Bacon's *Scientia Experimentalis*: A Review of Bacon's Diverse Treatises 12
JEREMIAH HACKETT

2 The Division of the Sciences in *Opus maius* Part IV, 1, 2: Roger Bacon's Sources and Their Transformation 42
ALEXANDER FIDORA

3 The Science of Weights: Jordan of Nemore's Influence on Roger Bacon's Scientific Method 55
YAEL KEDAR

4 Roger Bacon and John Pecham on Arithmetic 72
AURÉLIEN ROBERT

5 On Vision: Roger Bacon's *Opus maius* (V.1–2) from the Perspective of Alhazen's *Optics* (I–II) 94
NADER EL-BIZRI

6 Bacon on Mathematics, *Perspectiva,* and Astrology: The "Unnamed Master" Revisited 109
H DARREL RUTKIN

7 Knowability and Power of *Ars*: Roger Bacon's Utopian Science as Anti-Magic 130
NICOLAS WEILL-PAROT

8 Alchemy and Pharmacology in the *Opus maius* 147
MEAGAN S. ALLEN

9 Roger Bacon and Secrecy: Alchemical Techniques of Deception and the Doctor Mirabilis 165
WILLIAM R. NEWMAN

Index *183*

Contributors

Meagan S. Allen is a historian of science, specialising in the medical alchemy of the later Middle Ages. Her research interests lie at the intersection of alchemy, pharmacology, and theology, especially in the writings of the 13th century polymath Roger Bacon. She is the author of *Roger Bacon and the Incorruptible Human: Alchemy, Pharmacology, and the Desire to Prolong Life* (2023), which examines Bacon's desire to use alchemy to create a medicine that would extend human life by centuries. Allen is currently a Visiting Assistant Professor in the Department of the History of Science and Technology at Johns Hopkins University.

Nader El-Bizri is the Dean of the College of Arts, Humanities, and Social Sciences at the University of Sharjah. Previous affiliations include the universities of Cambridge, Durham, Zurich, Harvard, Nottingham, Lincoln, the London Consortium, the American University of Beirut, and the Centre National de la Recherche Scientifique. He served as a Referee to the Leverhulme Trust, the British Academy, the Arts and Humanities Research Council, the European Union Paris-Region Postdoctoral Fellowship Programs, and the Australian Academy of the Humanities. He has advised and consulted for many museums, including the Aga Khan Trust for Culture in Geneva.

Alexander Fidora is an ICREA Research Professor in the Department of Ancient and Medieval Studies of the Universitat Autònoma de Barcelona and a member of the Academia Europaea. His research focuses on medieval philosophy, particularly on epistemology, and the intercultural dimensions of medieval thought, with more than 40 books published and over 150 articles. He has directed the ERC Starting grant "Latin into Hebrew" (2008–2012) and the ERC Consolidator project "The Latin Talmud" (2014–2019). Previous appointments include Visiting Professor at Saint Louis University, Universidad Panamericana in Mexico, University of Erlangen-Nuremberg, and University of Pennsylvania.

Jeremiah Hackett is a Professor Emeritus of Philosophy at the University of South Carolina, where he was Chair of the Department of Philosophy. He has published on the philosophy of Roger Bacon and is the President of the

Roger Bacon Research Society. Recently, he and Thomas S. Maloney have completed a translation with an introduction and notes to Roger Bacon's *Philosophia Moralis* (2022). Hackett's current work on Roger Bacon concerns the different versions of the *Perspectiva* and the different versions of *De scientia experimentali*. He is currently preparing an edition of the first version of Bacon's *Perspectiva*.

Yael Kedar is a Senior Lecturer at Tel Hai College. She has published numerous papers on medieval science and philosophy, focusing mainly on Roger Bacon and Robert Grosseteste. She edited (with G. Hon) a special issue of *Studies in History and Philosophy of Science*, titled "Law and Order: The Status of Natural Regularities before the Scientific Revolution" (2020), (with J. Hackett) a special issue of *Revista Espanola de Filosofia Medieval*, titled "The Philosophy of Roger Bacon" (2021), and a book (with N. Polloni), titled *The Philosophy and Science of Roger Bacon – Studies in Honour of Jeremiah Hackett* (Routledge, 2021).

William R. Newman is a Professor in the Department of History and Philosophy of Science and Medicine at Indiana University. He has published 12 books, mostly dealing with the history of alchemy, including his 2019 work, *Newton the Alchemist: Science, Enigma, and the Quest for Nature's "Secret Fire"* (Princeton, 2019). Newman's work has received numerous awards, including a Guggenheim fellowship, the Pfizer Prize, and membership in the Institute for Advanced Study. He is currently General Editor of the *Chymistry of Isaac Newton* project (www.chymistry.org).

Aurélien Robert is the Directeur de recherche au CNRS and Président de la section 35 du Comité national de la recherche scientifique. He specialises in the history of medieval philosophy and science and has published several works on atomistic theories of the continuum, the reception of Epicurus in the Middle Ages, and the relationship between philosophy and medicine. Among his recent publications: *Epicure aux enfers. Hérésié, athéisme et hédonisme au Moyen Âge* (Paris, 2021), *Frontières des savoirs en Italie à l'époque des premières universités* (with J. Chandelier, Rome, 2005), and *Atomism in Late Medieval Philosophy and Theology* (with ch. Grellard, Leiden-Boston, 2009).

H Darrel Rutkin is a historian of science and philosophy specialising in the history of Medieval, Renaissance, and early modern astrology, cairca 1250–1800. His work focuses on astrology's numerous relationships to science, theology, and magic within their relevant contexts over the longue durée. Among numerous other publications, he has recently completed volume one of his first monograph, *Sapientia Astrologica: Astrology, Magic and Natural Knowledge*, circa 1250–1800, which is entitled *Medieval Structures (1250–1500): Conceptual, Institutional, Socio-Political, Religious and Cultural* (2019). He also recently co-edited with Denis J.-J. Robichaud, Marsilio Ficino's Cosmology: Sources and Reception, Bruniana e Campanelliana 26 (2020/2): 373–458.

Nicolas Weill-Parot is a Professor at the École pratique des Hautes Études, where his research deals with rationality in medieval science. After studying the relationships between science and magic, his focus turned to the internal borders of science: occult properties, magnetic attraction, abhorrence of a vacuum, and scientific utopia. He focuses in particular on Latin medieval commentaries on Aristotle's *Physics*.

The *Opus maius*

A Plea for Reform

Meagan S. Allen

In the first half of the 1260s, the English Franciscan and polymath Roger Bacon (ca. 1220–1292) was worried. Christendom was in danger. Seven official crusades – and many more unofficial ones – had left many desiring more effective (and less violent) means of conversion. Farther east, the Mongol Empire was growing larger, threatening to encroach on Christian territories. And above all, the fear that the Antichrist would soon appear remained ever present.[1]

Latin Christendom needed to be revitalised. But this renovation would not come from military supremacy: a final crusade to conquer the Holy Lands or an epic battle against the Mongol Khanate to spread Christianity throughout Asia. Rather, it would come about through science and the reformation of the universities. Through the study of the sciences, namely, languages, optics, mathematics, and experimental science, wise individuals would be able to attain divine wisdom. Bacon believed these had been neglected in the university curricula, and it was for this reason that he penned the *Opus maius* to Pope Clement IV in 1267, pleading that the pontiff use his power to champion their study. With a solid foundation in the sciences, one could learn about philosophy, which in turn would be applied to theology. For Bacon, philosophy was useless, unless dedicated to God, at which point it became an indispensable tool for the good Christian and the Church as a whole. It is philosophy's – and science's – utility to theology that makes it not only worthy of study, but an absolute necessity.[2]

As one of Bacon's most important works, there have been many studies on why – and how – Bacon chose to write the *Opus maius* for Clement IV. This volume explores the individual sciences found within the *Opus maius*, and places them in the context of thirteenth-century scientific activity. Bacon has long been considered a forerunner of modern science – anticipating contemporary methodology by several centuries. These essays will demonstrate, however, that while unique in some ways, Bacon was very much a product of his time. More importantly, they show that one does not have to prefigure modernity in order to be interesting or worthy of study.

In order to understand the context in which Bacon is writing, it is necessary to understand something of the man himself. The details of his early life are scarce, though he seems to have been born to a wealthy family sometime

DOI: 10.4324/9781003405702-1

around 1220, possibly in Ilchester. His life can be divided into two periods, the division being when he joined the Franciscan Order circa 1257. Around 1240, Bacon received an MA from either the University of Oxford or the University of Paris. Whether he started his degree at Oxford and finished it at Paris, or moved to Paris soon after completing the degree is unknown, but we do know that by the early 1240s, Bacon was in Paris. His firsthand accounts of seeing Alexander of Hales, who died in 1245, undoubtedly place him in Paris at this time. Once in Paris, Bacon took up a position as a lecturer in the Faculty of Arts at the University, where he spoke on Aristotle's *Libri naturales*. At some point between 1247 and 1250, Bacon resigned from his position in the Faculty of Arts and left Paris for Oxford, where he makes references to figures such as Adam Marsh and Thomas of Wales. However, he continued to travel back and forth between Paris and Oxford for some time. What he was doing during this time is uncertain – he was without a doubt engaging in his own scientific experiments – he tells us this himself in his *Opus tertium* – but whether he took up a teaching position at Oxford is unknown.[3]

Around the year 1257, Bacon joined the Franciscan Order, probably in Oxford.[4] He leaves no explicit rationale for this decision, although a number of theories have been proposed, including his admiration for Robert Grosseteste (d. 1253).[5] Much like his life before joining the order, Bacon seems to have split his time between Oxford and Paris, albeit this time with the Franciscan communities in each city. He died in 1292 and was buried in the Franciscan cemetery in Oxford.

Both periods of Bacon's life are marked by intense periods of writing – he is the author of dozens of known works on subjects ranging from optics, perspective, alchemy, and medicine to grammar and moral philosophy. However, he is arguably best remembered for his *Opus maius*, sent to Clement IV – his great plea for reform. The *Opus maius*, as the authors of this volume demonstrate, was a massive undertaking. The final form, as sent to Clement IV, contains seven sections of unequal length, which are themselves further divided into distinctions and chapters. The contents are as follows:[6]

Part 1: The Four Causes of Human Error.
Part 2: The Relationship Between Philosophy and Theology.
Part 3: The Importance of the Study of Foreign Languages.
Part 4: The Importance of Learning Mathematics; Geography.
Part 5: Optics and Perspective.
Part 6: Experimental Science.
Part 7: Moral Philosophy.

The most complete critical edition of Parts 1–6 and the beginning of Part 7, completed in 1900 by John Henry Bridges, is over 800 pages in length. Each section is devoted to how a particular science can – and should – be used in the service of theology. This was Bacon's goal: to convince his patron that the previously neglected sciences needed to be re-engaged to strengthen the Church. Not only did the *Opus maius* demonstrate the necessity of the sciences to philosophy and theology, it also demonstrated – through their connection to moral philosophy – the unity of the sciences.[7]

In writing the *Opus maius*, Bacon acted as a spokesman for the sciences. Rather than viewing them with suspicion or questioning their place in a Christian education, Bacon aimed to show that through their study, they could help attain divine wisdom.[8] In his prefatory letter to Clement IV, Bacon explains his rationale, writing:

> The Vicar of the Savior may command that everything useful in the sciences and languages and everything necessary for the wisdom of God be made available for the study of wisdom. For there are innumerable things not present in the Church at the moment, which would be of the greatest benefit and glory to her, as will be clearly shown in the works which I am sending to you.[9]

At another point in the letter, Bacon likens the sciences to the wood and stones of buildings. Even though the buildings have not yet been constructed, the materials – that is, the powers of the sciences – are present. Bacon will show how to use these powers to construct the buildings needed for wisdom. And it was especially pertinent that he send the *Opus maius* now, Bacon elaborates, for:

> The marvels of the arts and sciences are currently subject to great difficulties, especially in the times that are approaching the days of Antichrist and his followers, towards which he will be furious and angry, so that the study of wisdom will oftentimes be confused.[10]

The danger of the Antichrist is always present in Bacon's mind, and the Church must be prepared.

The persuasive nature of the *Opus maius* is consistent throughout. As others have pointed out, the *Opus maius* is not an encyclopedic work. It is not a failed version of the *Speculum* genre so popular in the thirteenth century. Rather, it is a *persuasio praeambula*.[11] Its main purpose is to persuade Clement IV to act. And who better to act on behalf of the sciences than the Vicar of Christ himself? Bacon certainly cannot enact the changes himself – he is a humble Franciscan with no money or status. The Pope, however, has authority inherent in his position, which Bacon wishes to utilise. The multitudes are easily swayed by those in power, especially if they are wise and righteous. Thus, the order and the happiness of the whole world depend on the wisdom and knowledge of the head of the Church.[12] By speaking to Clement IV, Bacon is speaking to all of Christendom.

Not only will the people allow themselves to be led by one who is wise, but God has also made this an excellent time for reform. While the previous popes were not able to devote time and energy to the study of the sciences – through no fault of their own – Clement IV has been given a rare opportunity of peace by God. By the authority of Clement's virtue, Bacon says:

> the flag of triumph has been erected above heaven, the sword has been withdrawn, the hostile parties have been thrown into hell, the peace of the church has been restored and the reign of joy has been granted to all believers.[13]

But it is not enough that peace has been attained; this period of peace must be capitalised upon with "the salutary promotion of scientific activity."[14] It is then that the unbelievers will be converted, the Antichrist thwarted, and the church restored.

Despite the fact that his prefatory letter addressed to Clement IV describes why it is important that this campaign be sponsored by a pope, Bacon was in contact with Clement IV, formerly Cardinal Guy de Foulques, before his ascension to the Papacy. Sometime before 1265, Bacon, through Foulques' clerk Raymond of Laon, proposed some writings on science and moral philosophy. Laon, mistakenly believing that Bacon had already completed said works, passed the message on to Foulques, who in turn sent word to Bacon, asking to receive copies. Bacon was unable to receive permission from his superiors to write, nor did he have the funds to secure supplies, so Foulques' request went unanswered.[15]

Unable to fund such a project on his own, Bacon began a desperate attempt to raise money. He first attempted to contact his wealthy family in England. As royalists, they were in dire circumstances during the Second Barons' War (1264–1267), and he received no reply to his plea. Even after the death of Simon de Montfort in August 1265, Bacon's family seems not to have recovered, and even in 1267 he complains that he has not heard from his brother "to this day."[16]

Like his family, Bacon's request from his friends proved equally fruitless. But not all hope was lost. In February of 1265, Foulques was elected Pope Clement IV, and Bacon saw a new opportunity for patronage. He again wrote to Foulques, now as Clement IV, through the English envoy William Bonecor. Not only had Clement IV already shown interest in Bacon's project, but he also now would have the funds to sponsor it as Pope. Additionally, while a mandate from a Cardinal would do little to impress his superiors, Bacon hoped that a mandate for writings from the Pope himself would persuade his superiors to allow him to embark on such a project. Fortunately for Bacon, Clement IV was indeed still interested in his project. However, in Clement's second mandate to Bacon, dated 22 June 1266, Clement IV makes it clear that Bacon must keep this project secret. He cannot tell his superiors what he is doing and why he needs time to write or conduct experiments, or the money to purchase supplies and copyists. There is also no money promised for the project. Bacon is once again in the position of having to pay for everything himself, under strict secrecy, and to do it all as soon as possible.[17]

Despite the obstacles, Bacon set to work. Initially, Bacon intended to write a larger, more comprehensive statement on science and philosophy – a *scriptum principale.* However, Bacon soon realised that this project was impossible for one person to undertake, for it required specialised knowledge in many areas.[18] He abandoned the *scriptum principale* in favour of a *persuasio*, the work which eventually came to be known as the *Opus maius*.[19] This work, as we have seen, was a hortatory piece, meant to demonstrate the utility of the sciences to the Church. As for the composition of the *persuasio*, Bacon found it much easier than producing a *scriptum principale*. Rather than writing a new work from scratch, he collected many of his previous writings on the sciences into one large text, supplying them with a unifying framework.[20]

Because much of the work had already been written in other forms, the turnaround for the *Opus maius* was fairly short. Sometime between the end of 1267 and the beginning of 1268, Bacon sent the *Opus maius* to Clement IV by way of his favourite pupil, a young man named John. He also included several supplemental works, including the *Opus minus*, some works on optics and perspective, and the introductory letter to Clement IV. Unfortunately for both Bacon and Clement IV, however, we do not know if his patron ever had the chance to read these works. Clement IV died on November 29, 1268, having left no record of ever receiving or reading the *Opus maius* or any of its supplements.[21]

Today, the *Opus maius* is considered the first of three parts of Bacon's *Opera*, along with the shorter *Opus minus* and *Opus tertium*. Though various timelines have been proposed for their relative compositions, it is probable that they were completed between Clement IV's second mandate of June 1266 and the end of 1267.[22] The question must then be asked: if Bacon wrote the *Opus maius*, why did he feel the need to supplement it with two additional works? The explanation is given to us by Bacon himself – he is worried about the dangers of the road and wants to prepare a summary of the work, in case the original should get lost. If something should happen to either the *Opus maius* or John en route to the Papal Curia, the *Opus minus* will serve as a summary. Additionally, Bacon says, there are some things that he felt he left unclear or neglected entirely in the *Opus maius*, and he hopes that his further discussion in the *Opus minus* will elucidate them further.[23]

The chapters in this volume place the *Opus maius* within the context of the *Opera*, Bacon's oeuvre as a whole, and the broader scientific culture of the thirteenth century. Each author's disciplinary identity informs their chapter, leading to the rich variety of methodological approaches in this volume. In the first chapter, Jeremiah Hackett explores Bacon's conception of *scientia experimentalis*. According to Bacon, *scientia experimentalis* has three prerogatives: to verify the conclusions of the other sciences, especially those that are purely rational or based on imperfect experiences; to gain knowledge of the future; and to construct new technologies. By placing Bacon's *scientia experimentalis* within the context of the thirteenth century and his sources, namely, Ptolemy's *Quadripartitum*, Hackett demonstrates that Bacon derives two ways of gaining knowledge: logical argument and experience. Logic can be reduced to mathematics, and experience can further be divided into two kinds: that provided by external senses and that provided by internal illumination. Internal illumination has seven stages, which are connected to both moral truth and internal religious experience, namely: objectivity gained through scientific knowledge; knowledge of moral virtues; gifts of the Holy Spirit; beatitudes; spiritual senses; spiritual fruits; and mystical raptures. Ultimately, Hackett demonstrates that in Bacon's model of the sciences, *scientia experimentalis* is more essential to theology than any other science because it is the closest to morals.

In Chapter 2, Alexander Fidora explains Bacon's division of the sciences in Part 4 of the *Opus maius*. Though Bacon's division of the sciences was created within the framework of influential texts, such as al-Farabi's *De scientiis*, Bacon reinterpreted them by adding more emphasis to the moral aspects of philosophy. Focusing on Part 4, Distinction 1, Chapter 2, Fidora shows that Bacon divides philosophy into essential and accidental parts. The accidental parts derive from thirteenth-century Parisian philosophy, where they were defined as grammar, logic, and rhetoric. The essential parts, which can further be separated into mathematics, natural parts, and divine parts, come from Aristotle's classification of theoretical philosophy. But while early masters of philosophy identified these with Aristotle's speculative philosophy, Bacon includes practical and moral philosophy in this category as well. This went against the standard division of philosophy held by his contemporaries. As for the accidental parts of philosophy, Bacon directs their use towards moral philosophy (now a type of essential philosophy). Rhetoric and poetics serve as the link between accidental philosophy and moral philosophy.

Yael Kedar's chapter on Bacon's geometrical diagrams argues for a new source for his methodology: Jordan of Nemore. While previous scholars have demonstrated that Bacon relied on earlier authors, including Robert Grosseteste, Ptolemy, and Alhazen, Kedar shows that Bacon's search for underlying mathematical causes comes from Jordanus. Focusing on Part 4, Distinction 3, Chapter 16 ("The Science of Weights"), which deals with the motions of the elements, it becomes evident that Bacon was less interested in the science of weights itself than in using this study to prove that geometry is effective for the study of motion. Specifically, he wished to find geometric causes for motion through the use of diagrams that visually depict relationships between objects. Bacon's contribution to the study of weights is not in the form of any new theory but in the fact that he is able to apply the methodology laid out by Jordanus, which he would use in other demonstrations.

Further exploring Bacon's conception of mathematics, Chapter 4 asks why Bacon's understanding of arithmetic has tended to be overlooked in scholarship. Aurélien Robert argues that this is because Bacon himself considers arithmetic to be relatively unimportant compared to the other sciences of the quadrivium, because it does not give direct access to natural laws or the ontology of objects. Rather, it is an arbitrary means of numbering objects that have been identified by different methods. Robert demonstrates that Bacon's critical view of arithmetic largely arises from his rejection of atomism – both mathematical and metaphysical. This leads him to give priority to the continuous quantities of geometry over the discrete quantities of arithmetic, despite the tradition of placing arithmetic at the head of the quadrivium. Instead, Bacon argues that the quadrivium should be oriented towards perspective, which will show the structure of the world and the laws of nature. Arithmetic, on the other hand, is primarily useful for practical applications, such as verifying calculations and measurements of sacred texts. Finally, by comparing Bacon's approach to arithmetic to that of John Pecham, Robert shows that there is no one "Franciscan" way of approaching mathematics in the thirteenth century.

In Chapter 5, Nader El-Bizri compares Bacon's approach to perspective and vision to that of his main source, Alhazen's *Optics*. However, Bacon does not merely copy Alhazen's words uncritically; rather, he synthesises from a number of secondary sources, including Avicenna and Constantinus Africanus. After describing the anatomy of the eye, Bacon highlights the role of the common sense in the "cognitive aspects of visual perception." Emphasising the imagination and the recollection of memory in the interaction between vision and the mental faculties, Bacon divides the senses into two classes. The first class, which is external, comes through our sensory experience of the environment and activates the common sense in the anterior of the brain. The second class, which is internal, relates to the judgements of visual perception. This class associates an intentional meaning to an object, resulting in an emotion or reaction. In this way, Bacon is able to explain both the physiological process of vision, as well as the effects vision has on the viewer.

Chapter 6 concerns the relationship between mathematics and another science of the quadrivium – astronomy. After describing the relationship between astronomy-astrology and mathematics in Bacon's *Opus maius* and his edition of the Pseudo-Aristotelian *Secretum secretorum*, H. Darrel Rutkin argues that Bacon divides *astrologia* into two categories: *astronomia iudiciaria* and *astronomia operativa*. While *astronomia iudiciaria* is interpretive and knowledge-based, *astronomia operativa* is technological and concerns the creation of talismans to create effects in the sublunary world. Theologians have misunderstood and suspected this science, Bacon says, because there is a false type of mathematics, an illegitimate astrology, which is deterministic and invalidates free will. This, however, is entirely different from the true and legitimate forms of astrology that do not undermine free will. With this understanding of Bacon's astrology in mind, Rutkin proceeds to revisit the question of the "Unnamed Master" in the *Opus tertium*. Who is the unnamed master who is ignorant of perspective and the subject of Bacon's ire? By comparing the doctrines found in Bacon's works with those of Albert the Great, Rutkin provides further proof that the "Unnamed

Master" is in fact Albert the Great, even if Bacon's criticism of Albert is unfair.

H. Darrel Rutkin's chapter demonstrates how Bacon is violently opposed to the concept of "magic," placing it in opposition to the legitimate science of astrology. In the following chapter, Nicolas Weill-Parot delves further into Bacon's concerns surrounding the fraudulent art of magic, this time comparing it to *scientia experimentalis*. Weill-Parot argues that Bacon conceives of *scientia experimentalis* as being a utopian science, that is, an extraordinary project in a scientific framework. Such a project is beyond the current limits of scientific capacity but remains theoretically possible. The distinction for Bacon lies in the fact that human art can not only equal nature but can in fact surpass it, allowing for all natural causes to be elucidated. It is this issue of secrecy in causes that distinguishes between *scientia experimentalis* (good) and magic (bad). While magic is fraudulent and fundamentally relies upon concealment and ignorance, the secrets of science are meant to be revealed as one progresses on the path towards knowledge. Ultimately, Weill-Parot argues that Bacon largely rejects the notion of occult action, that is, definitively unknown causes, in favour of secrets, which are only temporarily unknown but discoverable through *scientia experimentalis*.

The theme of secrecy continues through the final two chapters of this volume. In Chapter 8, Meagan S. Allen examines the unusual relationship between alchemy and pharmacology in the *Opus maius*. In Part 6 of the *Opus maius*, both alchemy and the *prolongatio vitae*, or extreme prolongation of life, are considered aspects of *scientia experimentalis*. Bacon tells numerous stories of individuals who have been able to prolong their lives by decades, even centuries, by ingesting substances that have had all of their elements balanced in the proper proportion. Bacon also provides a list of pharmaceuticals that he thinks are particularly efficacious for prolonging life. Alchemy, while also a form of *scientia experimentalis*, is not related to the *prolongatio vitae* and is concerned with the creation of purer forms of gold. However, this is a significant departure from Bacon's other writings on the *prolongatio vitae*, in which alchemy is the science most necessary for the prolonging of life, for it alone teaches how to balance the elements. As early as the *Opus minus* and *Opus tertium*, Bacon abandons the pharmaceuticals found in the *Opus maius* in favour of alchemically produced medicines. Allen argues that this was done intentionally, as a way of protecting the potentially dangerous information found in the *Opera*. Afraid that one manuscript will be stolen and the alchemical information will fall into the wrong hands, Bacon splits it up over the three parts of the *Opera* so that only his intended audience – Clement IV – will have access to the knowledge. The pharmaceuticals listed in the *Opus maius* act as mere placeholders for the more important alchemical medicines of the later works.

In the final chapter, William R. Newman characterises the five types of alchemical secrecy found in Bacon's writings: Decknamen – cover-names for alchemical ingredients; syncope – the omission of operational stages or ingredients; parathesis – the unnecessary repetition of stages or ingredients, often with different names; graduated iteration – the use of one term to mean multiple things; and *dispersa intentio* – the breaking up of theories or steps throughout a text (or multiple texts). The use of secrecy in alchemical works was not novel to Bacon; *dispersa intentio* was a characteristic of the works of the Arabic Jabirian corpus and can be found in early Latin translations of alchemical works that may have been available to him, including Gerard of Cremona's *Liber de septuaginta*. It was from this alchemical tradition that Bacon developed his view of the necessity of secrecy. A larger examination of Bacon's corpus shows that the use of literary secrecy is not limited to the *Opera*. Newman concludes his chapter with an examination of Bacon's *Epistola de secretis artis et naturae et de nullitate magiae*, the eighth chapter of which is devoted entirely to the various techniques of literary concealment. Literary concealment, especially in the context of alchemy, was thus a fundamental characteristic of Bacon's writings. However, it is not limited to alchemy, and Bacon considers it important for any secrets of the wise that have the potential to be perverted by the masses.

Each chapter in this volume was originally a lecture presented to the Roger Bacon Research Society, which seeks to study Roger Bacon's thought in the context of the broader milieu in which he was writing. Bacon has often been painted as a lone visionary in a century devoid of science. As our chapters and our society aim to show, this is not the case. Bacon did not invent novel theories in every scientific discipline in which he worked, and there certainly was no dearth of science in the Middle Ages. But this does not diminish the importance of Bacon's work. Bacon's genius often lay elsewhere – he was an exceptional synthesiser of various, often apparently incommensurable, theories. He held appreciation for sciences that others deemed useless. And above all, he searched for practical applications for the sciences that would have positive, lasting effects in his own world. We hope these chapters will provide valuable new insights into the sciences contained in Bacon's *Opus maius*, a truly great persuasive work written on behalf of scientific and educational reform.

Notes

1 For more on Bacon and the Mongol Court, see Maloney and Hackett, "Introduction," 19.

2 Bacon, *Letter to Pope Clement IV*, 153. This is a translation of what has been referred to as the "Gasquet Fragment." Easton identifies this work as either a prefatory letter to the *Opus maius* or part of the *Opus minus*. Either way, it seems certain that it was sent to the Pope with some part of the *Opera*, although there is no record of the Pope having read it. See Easton, *Roger Bacon*, 145.

3 Lindberg, "Introduction," xvii–iii. Lindberg thinks that Bacon was one of the early lecturers on Aristotle after the recurring bans in Paris in 1210, 1215, and 1231. Bacon was certainly in Paris in 1251, for he gives a firsthand account of the uprising of the Pastoreaux rebels.

4 Lindberg, "Introduction," xx. The dating of 1257 comes from a passage in the *Opus tertium*, where Bacon says that he has been in exile from his previous fame acquired from studies for 10 years.
5 Crowley, *Roger Bacon,* Chapter 1.
6 Bacon, *Opus maius*, Parts 1–7.
7 See Easton, *Roger Bacon*, 146–166; and Maloney and Hackett, "Introduction," 17.
8 Bacon, *Letter to Pope Clement IV*, 150.
9 Bacon, *Letter to Pope Clement IV*, 161.
10 Bacon, *Letter to Pope Clement IV*, 145.
11 Bacon refers to the *Opus maius* in these terms. See also Bridges, *The Life and Work of Roger Bacon*, 138–139.
12 Bacon, *Letter to Pope Clement IV*, 146–147; 155.
13 Bacon, *Letter to Pope Clement IV*, 146.
14 Bacon, *Letter to Pope Clement IV*, 146.
15 Various accounts of this exchange occur in Bacon's letter to Clement IV, as well as Easton, *Roger Bacon,* 144–166; Crowley, *Roger Bacon*, 35; Lindberg, "Introduction," xxii; and Brewer, *Fr. Rogeri Bacon OQHI,* xvi–ii. Bacon's difficulty in writing would have stemmed from the censorship resulting from the 1260 Chapter of Narbonne. This prohibited the composition of books without the licence of the Minister-General of the Franciscan order, as well as direct communication with the Papal Curia.
16 Easton, *Roger Bacon*, 147.
17 See Easton, *Roger Bacon,* 146–166; Bridges, *The Life and Work of Roger Bacon*, 13–14; Brewer, *Fr. Rogeri Bacon OQHI*, xvii; Lindberg, "Introduction," xxii.
18 Bacon, *Letter to Pope Clement IV*, 150.
19 Lindberg, "Introduction," xxiv–xxv. Bacon abandoned the *scriptum principale* after Epiphany in 1267.
20 Easton, *Roger Bacon,* 161.
21 Lindberg, "Introduction," xxv. It is certain that the works made it to the Papal Curia – the Polish mathematician Witelo (d. 1278) was aware of their contents.
22 In the *Opus tertium*, Bacon speaks of 1267 as the current year. For the different proposed timelines and their critiques, see: Brewer, *Fr. Rogeri Bacon OQHI*, xlvi; Lindberg, "Introduction," xxv; and Easton, *Roger Bacon*, 146–166. Easton also provides a thorough review of the timelines proposed by Mandonnet and Thorndike. While the *Opus maius* and *Opus minus* were sent to Clement IV, it is likely that the *Opus tertium* was never dispatched.
23 Easton, *Roger Bacon,* 156–157. Statements to this effect can be found in both the *Opus minus* and the *Opus tertium.*

Bibliography

Primary Sources

Roger Bacon. *Letter to Pope Clement IV*. Translated by Nikolaus Egel, *Revista Española de Filosofía Medieval*, 27/2 (2020): 143–174.

Roger Bacon. *Moral Philosophy*. Translated with Introduction and Notes by Thomas S. Maloney and Jeremiah Hackett. St. Bonaventure, NY: Franciscan Institute Press, 2022.

Roger Bacon. *Opus majus*. Edited by John Henry Bridges. 3 vols. Oxford and Edinburgh Clarendon, 1897–1900.

Roger Bacon. *Opus tertium*. Edited by John S. Brewer. *Fr. Rogeri Bacon Opera Quaedam Hactenus Inedita, vol. I: Containing the Opus Tertium, Opus Minus, and Compendium Philosophiae*. London: Longman, Green, Longman, and Roberts, 1859.

Secondary Sources

Bridges, John Henry. *The Life and Work of Roger Bacon.* London: Williams and Norgate, 1914.
Crowley, Theodore. *Roger Bacon: The Problem of the Soul in His Philosophical Commentaries.* Louvain: Éditions de l'Institut supérieur de philosophie, 1950.
Easton, Stewart C. *Roger Bacon and His Search for a Universal Science.* New York: Columbia University Press, 1952.
Lindberg, David C. *Roger Bacon's Philosophy of Nature.* Oxford: Oxford University Press, 1983.

1 Roger Bacon's *Scientia Experimentalis*

A Review of Bacon's Diverse Treatises

Jeremiah Hackett

1.1 Introduction

By the early fourteenth century, Roger Bacon had emerged as an icon of English medical learning. Francis Bacon in the early modern period presented Roger Bacon as an exception to Scholastic dialectical discourse; he is the true experimentalist who seeks out the secrets of nature. This latter image of Roger Bacon as an exception to Scholastic education came to be the norm for writers on the History of Science in nineteenth-century Britain. By the early twentieth century, it was not unusual to portray the *quondam Magister in Artibus* and subsequently Franciscan friar, Roger Bacon, as "the founder of modern science." Two eminent historians of science, Lynn Thorndike and Pierre Duhem, who knew Roger Bacon's texts and context better than the nineteenth century historians, deflated these exaggerated claims of the modernity of Bacon's science. Nevertheless, the thesis that both Robert Grosseteste and Roger Bacon outlined the *qualitative but not quantitative* dimensions of the modern scientific method was carefully argued in 1953 by A.C. Crombie. Crombie's thesis received a strong rebuke from the historian of Galilean Science, Alexander Koyre.[1]

Contemporary writers have placed Bacon's understanding of experience and experiment firmly back in their medieval context. A judicious study of prerogative one of Bacon's science of experience in relation to *Perspectiva* was provided by David C. Lindberg in 1997.[2] The present author reviewed the main sources of Bacon's *scientia experimentalis*.[3] Recently, some very significant studies on specific aspects of Bacon's science have been published. First, Cecilia Panti has reviewed in detail the relation of Grosseteste to Bacon on the theory of the rainbow. She has demonstrated how, for both Grosseteste and Bacon, the case of the rainbow, halo, and related phenomena is a *locus classicus* for the applications of *Perspectiva* to individual phenomena in nature.[4] Second, Bacon's example of the prolongation of life through medicine as part of the second prerogative has been masterfully outlined in a new book study by Meagan Allen.[5] Third, P. Schramm's thesis on universal laws in Bacon's science has been taken up and developed by both Yael Kedar and Giora Hon.[6]

DOI: 10.4324/9781003405702-2

The main result of modern and contemporary research on Roger Bacon's science is the recognition that Roger Bacon was not *the founder* of modern science; he was a thirteenth-century university teacher of mathematics and science at Oxford and Paris. Later, circa 1257–1274, as a Franciscan friar in the studium in Paris, Bacon was a scholar of the Latin Biblical text with reference to Hebrew and Greek original sources. He also tutored students in languages, logic, the sciences, and moral philosophy. Bacon wrote the first comprehensive work on *Perspectiva* in the Latin West (ca. 1262–1268). This latter work gained him significant fame, placing him among the first perspectivists, including John Pecham and Witelo. Master Peter of Limoges refers to him as an *Auctor*.[7] And it is as the first Latin authority on optics that Bacon comes to be known in the later Middle Ages, Renaissance, and early modern times.

And yet, there is still a need for a more comprehensive review of the main treatises where Bacon treats experience and science. The present chapter provides a review of the relevant Baconian texts. However, reference to Bacon's account of the rainbow and halo will be set aside for another occasion. The focus here will be on Bacon's concept of experience and the three prerogatives of the science of experience as presented in his later works.

1.2 The Technical Expression *Scientia Experimentalis* in Roger Bacon: Its Source and Context

In his discussion of the magical arts of geomancy, aeromancy, hydromancy, and pyromancy in Chapter 5 of his Introduction to his edition of the Pseudo-Aristotelian *Secretum secretorum*, Bacon states that there are other such names that are not parts of the art of magic but are parts of philosophy and experimental science.[8] He states:

> It shall be known that the above-mentioned geomancy is magic, as is a certain kind of hydromancy and pyromancy, about which matters there are books of magicians. But other books with the same names, but not having the same meaning or content, are parts of philosophy and experimental science. For according to Ptolemy [Pseudo] in the book *On the disposition of the sphere*, which under another title is named *The Introduction to the Almagest*, the part of philosophy dealing with knowledge of the future according to possibility, as is said, is two-fold: one is Natural Astronomy, which judges about natural things in this lower world, as is said, that is a middle way between the necessary and the impossible, as Ptolemy [Pseudo] says in the *Centiloquium* and Ptolemy in his *Quadripartitum* such that astronomers do not impose necessity on freedom of choice nor on contingent events as has been explained above.
>
> But there is another science which considers future alterations in this inferior world, and in the aforementioned book it is called *scientia experimentalis* by Ptolemy [Pseudo] about which as Ptolemy [Pseudo]

> states Noble Aristotle is based and many crowds of the lords of judicial astronomy whom Aratus the Philosopher augmented.[9]

And this science considers those things which appear on earth through signs in animals and certain terrestrial and aqueous things, and the impressions that appear in the air and in fire, that is, in the sphere of fire, through which philosophers judge about wind and rain and dryness and heat and cold and humours, pestilence and putrefaction of animals, and in water and humans and all living things. Bacon discusses the fish in the sea, the sound of the wind moving through the forest, and the various properties of animals as certain signs of the changes of the air and of inferior things.

Similarly, those things which appear in the air, including inflamed impressions or inflamed vapours such as Assub ascending or descending, which the common people think are falling and descending stars. These are influential vapours which signify the dryness of inferior things, as is clear from the end of Ptolemy's [Pseudo] *Centiloquium.*[10] And many illuminated and influential vapours appear in diverse figures concerning which Aristotle speaks in the *Meteorologies,* Book One, such as a flaming constellation, the serpents, and the figure of the pyramid, and many other such things.[11] Also, the *rainbow and halo or Alileti,* which is the coloured circle around the sun, moon, and the stars, and the coloured perpendicular virga, signify many things in the changes of the air.[12] Bacon holds that comets, which are called stellar comets, signify abundantly. This, he says, can be seen in his own times with the appearance of the great comet of 1264, which led to disturbances of the air in all known regions. He notes the social and moral disturbances that arose from the appearance of the great comet. Bacon continues with a discussion of the eclipse of the sun and moon.

> All these and similar phenomena are the effects of the planets and stars and by Ptolemy [Pseudo] are called "secondary stars" because they are effective in the second place in respect of their cause. Concerning this Ptolemy [Pseudo] says in the *Centiloquium* that the soul of the wise person judges more certainly by the secondary stars than by the stars. And so, it is stated in the *Liber de dispositione sphere* that noble Aristotle and the Astronomers based themselves on this science more than on astronomy.[13]

What follows here is very important. According to Bacon, the matters about which philosophy and *scientia experimentalis* deal are totally different from the uses of earth, fire, water, and air in the books of magicians. Experimental sciences are natural sciences. They study the signs of the future in these natural phenomena in a natural scientific manner, just as they study comets and eclipses in the sphere of fire. Aristotle, in his book *Meteorologies,* proves that the study of earth, fire, water, and air are true parts of philosophy. But the Latin translator of the *Secretum secretorum* did not understand the proper

Latin scientific terminology, and so he employed words from the books of the magicians which seemed like the words of certain true sciences.

That Bacon bases his defence of experimental astrology on Ptolemy's *Quadripartitum* is not so obvious from the brief introduction to the *Secretum secretorum*. Still, these Pseudo-Ptolemaic texts presuppose and comment on Ptolemy's *Quadripartitum*. The importance of the *Quadripartitum* for Bacon's justification of his *scientia experimentalis* becomes very clear when one examines his distinction between true mathematics and false [or magical] mathematics in the *Opus maius*, Part 4. This latter text is important for his defence of an art and science of nature in *Opus maius*, Part 6, and related texts.

In *Opus maius*, Part 4, Bacon acknowledges that Christian scholars do not attack Ptolemy, Avicenna, and many other writers because of "sacrifices, conjurations, incantations, and characters." Their main concern is with the infallibility of judgements and necessitarianism in events. Bacon begins his defence of judicial astrology with reference to [Verbum 1 and the Expositio]: "Ptolemy states that the astrologer ought to speak in universal terms and not about specific things. [This is so] because this science proceeds according to demonstrative proof and opinion."[14] Bacon provides examples of the application of astrology by reference to [Verbum 59]. He then turns to Ptolemy's *Quadripartitum*.[15] Here, Bacon provides a summary of Book 1, Chapters 1–3 of this latter text.[16] Bacon cites this work by Ptolemy to argue that the astronomer does not provide certain judgements for each individual case. Instead, he should make universal claims about the possible outcomes situated between the necessary and the impossible. The astrologer cannot give a final judgement. Bacon concludes:

> For this reason, Avicenna, who completed the works of Ptolemy, as he states in the prologue to the *Book of Sufficiency* demonstrates in the tenth book of his *Metaphysics*, that an astrologer cannot claim certainty in all cases nor should he, because of the instability of generated and corruptible matter, which does not always conform to the heavenly forces...as Messahalac states, taking the magnet as an example.[17]

It should be noted that Bacon's concept of experience in the sciences is not only influenced by Ptolemy's *Quadripartitum* and the two Pseudo-Ptolemaic works mentioned above. Bacon was also significantly influenced by the first chapter of Ptolemy's fundamental work, the *Almagest*.[18] In that work, Ptolemy gives a decidedly empirical and especially mathematical turn to Aristotle's concerns.[19] Mathematics alone provides the highest standards of proof. Both physics and theology are matters of conjecture. In this regard, mathematics can aid them as an instrument. Indeed, the applications of mathematics in *Opus maius*, Part 4 to physics, the calendar, geography, theology, and astrology can be seen as Bacon's outline of Ptolemy's general project. In his *Communia mathematica*, Bacon indicates his concerns with the importance of the first chapter of the *Almagest*. Bacon's first reference deals with the division given by Ptolemy (taken from Aristotle's *Metaphysics*) of the essential parts of philosophy into mathematics, natural philosophy, and theology. "And as Ptolemy teaches in the first chapter of the *Almagest* mathematics is of importance for an understanding of the other two essential modes of science [philosophy]."[20] Bacon sees Boethius in his *Arithmetic* as one taking up Ptolemy on the application of mathematics in reference to civil life, and he finds this supported in Aristotle and his commentators on civil science. Bacon finds further support for the application of mathematics in Alfarabi's *De scientiis* and in Aristotle's *Poetics*.[21]

Still, it is not only the issue of mathematics that matters in Ptolemy's introduction to the *Almagest*. The *empeiria* of Aristotle is given a very specific experiential turn and is linked closely to mathematics. Another reference to the *Almagest* in Bacon's *Communia mathematica* concerns the parts of astronomy; I will return to this below in reference to astrology. The main point here is that for Bacon, all the ancient and medieval authorities cited above address a science of experience where mathematics and careful observation of phenomena are combined. Logical and mathematical arguments are essential for demonstration. Still, in matters dealing with the heavens and the Earth, verification by experiences is the only way we have to distinguish an art and science of nature from fraud, exaggeration, and the purely magical. But another important issue arises here. Bacon states that the experience mentioned in the *Centiloquium* is not only scientific experience; there is another, higher kind – internal experience (*ex se*). This is a kind of internally inspired knowledge.[22] The truly wise person must have both scientific knowledge and internal intuition. Bacon is talking about knowledge of future events.

1.3 Bacon's Acknowledgement of His Own Imitations in Experimental Science

Roger Bacon never made the unqualified claims about his experimental science that some modern interpreters have imposed on him. Although Bacon was a very learnt teacher of languages and sciences and a Biblical scholar, he was quite modest about the limitations of his own knowledge, particularly with respect to *perspectiva* and experimental science. I give two examples, one from the *Opus maius*, the other from the *Opus minus*. In the former work, writing on the first prerogative of his experimental science dealing with the rainbow and the halo, Bacon writes:

> Therefore, no [mere] discussion can give certainty in these matters, for the whole subject is dependent on experience. For this reason, I do not think that in this matter I have attained the whole truth, because I do not yet have the expertise in all things that are necessary for this science and because in this work I proceed by way of persuasion and exhibition of those things which ought to be required in the study of wisdom and

> not by the method of composing [a formal Treatise] on these sciences. And so, it does not pertain to me now to give an impossible verification, for I turn instead to the format of a rhetorical persuasion about the study of wisdom.[23]

Thus, Bacon was not writing a formal treatise on *Perspectiva* and *De Iride* such as one will later find in Witelo or Dietrich of Frieberg. That Bacon is aware of his own limitations in perspectival knowledge is also clear from the *Opus minus*. He comments:

> Clearly the things said about the rainbow and a coloured circle occupied me for a month before I was able to address them by other drawings and other experiences. And because the principal work that you seek requires these things, yet I did not satisfy myself nor the matter about which I am writing, especially since the rainbow and those circles rarely appear in the way that a method for how to certify them requires. And I am certain that none among the Latins except one person, who is the wisest of the Latins, could meet with approval in this matter. And here I estimate that I have met my match more than in all [other instances] to which I give attention, because of the wonderous experiences/observations that are required here not only during the day but also at night on behalf of the Moon's rainbow and the coloured circles around it.[24]

1.4 Bacons's Experiences in Science and the Trustworthy Experimenter (*Fidelis Experimentator*), the Wisest among the Latins: *Magister Petrus*

In the *Opus tertium* (ed. Brewer/Egel), Ch. XIII, Bacon addresses the role of *scientia experimentalis* and identifies the one who is, in his view, the best representative of this kind of science (ca. 1262–1268). Bacon presents this science as follows:

> Apart from these latter sciences there is one which is more perfect than all, which all other sciences serve and which in a wonderful way certifies all other sciences. And this science is named experimental science, which avoids mere arguments, since bare arguments do not certify the conclusions of arguments, even if the latter are strong, unless those arguments also include an experience of the conclusion, as I show in the treatise on this science.[25]

It is the task of the first prior-claim (prerogative) of this experimental science to certify the other sciences whose arguments are either purely rational or based on imperfect experiences. These are the truths proposed in natural philosophy and perspective concerning "the rainbow, the circles around the sun, moon, and the stars." Bacon states that he demonstrates the greatest power of this first prerogative of experimental science for the purpose of certification:

> Certain natural philosophers [commenting on] the *Meteorologies* of Aristotle and some experts on *Perspectiva* engage in vain in trying to certify these truths. And so I posit the roots of the experiments about these matters, which no Latin thinker except one is able to understand, namely Magister Peter. And this is the case also with regard to [the sciences] of all natural things. There is a second prior-claim (prerogative) for this science. It can verify noble truths in the terminology of other sciences which those sciences themselves cannot prove nor in any other way investigate. Such is the case with the prolongation of life and many other things. For this truth is in the terminology of medicine, but nowhere can the art of medicine extend itself to the verification of these truths. Nor is there anything of this kind in the books on medicine.[26]

Just as the Sea-Captain will order the carpenter to build a ship or the military general will order the armament maker to construct good arms, so the *Experimentator Fidelis* will know how to examine animals to discover new medical cures and will know how to control animals:

> The third prior claim (prerogative) of this experimental science does not use the terminology of the other sciences but uses them as servants. This prerogative has two tasks: one is knowledge of the future and the other is the construction and use of new works of art and nature.[27]

Here, Bacon repeats what he said in the *Opus maius*, the introduction to the *Secretum secretorum,* and related texts: this is a way of judgement about the future which is more certain than astronomy and is the way of Aristotle, many faithful philosophers, and lords of judicial astronomy.[28] The second part of this science deals with the very nature of experimental science. It teaches us about all the possibilities of this science.

Here in the *Opus tertium* and repeating material from *Opus maius*, Part 6, Bacon states that just as logic teaches the properties of sophistical argument so as to exclude it, this experimental science examines all the arts of magic so as to reprove them. It does this so that having reproved all falsity, only the truth of art and nature will be retained. The context of Bacon's demarcation of magic and an art and science of nature is the fact that the Canon Lawyers (the followers of Gratian) condemn both the good and the bad in natural science:

> And Gratian and many of his followers speak less well in this matter. For not everything which they teach should be condemned is a part of magic. And they condemn as false that of which they are ignorant. But

> the one who condemns something ought to know the actual conditions of what is condemned, and then falsehood can be condemned while the truth will always be preserved.[29]

Bacon then turns to what is, perhaps, his greatest example of a product of experimental science, namely the construction of a burning mirror.[30] While the production of such a mirror involves a knowledge of geometry (since a knowledge of figuration is required), an expert in geometry alone could not do this marvellous experimental work nor make use of it. Only the *Experimentator Fidelis* knows how to think through this great work of art. This faithful experimenter commands the geometer to provide the geometrical plan for the burning glass. The use of the mirror is the task of the experimenter; the provision of the shape is the work of the geometer. This experimental science is the queen of all the other sciences and is the goal of all speculation. As in the *Opus maius,* so here, Bacon advocates the military uses of such burning mirrors.[31] They may be of better assistance than single armies in holding off the enemies of Christendom such as the Mongols and the Saracens. What follows is a description of the *Experimentator Fidelis, Magister Petrus*, commonly identified as Master Peter Peregrinus of Maricourt. This attribution is found in the earliest extant manuscripts of the *Opus tertium. Master Peter Peregrinus of Maricourt* is, of course, the author of the first experimental treatise on the magnet and the new treatise on the astrolabe.[32] Bacon speaks of him as follows:

> I know but one who can be praised for works in this science. For he himself does not care about speeches and verbal quarrels but seeks out [the scientific] works of wisdom and finds his satisfaction in them. And so, since all are accustomed to see things like blind people, as the bat sees the light of the sun in the evening, he examines things in the fulness of light. Because of this, he is the Lord of Experiments. And so he knows the nature of things by means of experience, medicines, and alchemical matters, and all things in the heavens and on Earth. Moreover, he would show modesty if any lay person or old woman or knight or countryman knew something of which he was ignorant. For he is competent in all works of the foundry for metals, how to work with gold and silver and other metals and all minerals. And he knows all things that pertain to the military and to arms, and to the hunt. And he has examined all things to do with agriculture and the measurement of land and the works of countrymen. Also, he has considered all *experientiae* (experiences/experiments) of old women and casting of witchcraft and their songs and all aspects of magic, and similarly all the illusions and tricks of players of games so that nothing which should be known would be lacking to him to the point that he knows how to reprove all falsehood and magic. And so without him, it is impossible that philosophy would reach its completion or be treated usefully and

> with certitude. And just as no one is suitable for praise, neither does he calculate his own. For if he wished he could mix with Kings and Princes, and he would easily find those who would honor and enrich him. And if he wished to show off in Paris all that he knows about the works of wisdom, all the world would follow him. But through each of these ways, he would be impeded from the greatness of his experiments, in which he finds the greatest delight, thus he refuses all honors and riches, especially he would be able when he wished to reach riches by means of his wisdom.[33]

And so, we now can see that when Bacon speaks about his much-lauded experimental science, he is not speaking primarily about his own experimental work. He is acting as the spokesperson for a new paradigm of experiential knowledge, one quite distinct from the learning of the teachers of the arts at the universities. And he is specifically writing to the Pope on behalf of an important French experimentalist, namely, Magister Peter Peregrinus of Maricourt. Indeed, this may be Bacon's way of saying that Albert the Great was not the only one who examined the phenomena of nature and magic in his many journeys. More importantly, these and other remarks by Bacon tell us that he was part of a group which was concerned with making a strict distinction between the art and science of nature and the arts of magic. Perhaps, it could be argued that Bacon was also involved with Magister Peter and other scholars who sought to distinguish true books on the art and science of nature from books on the parts of magic.[34] Certainly, Bacon is presenting a resume of the experimental work of Magister Peter for Pope Clement IV's information and support.

1.5 The Introduction to Bacon's *De scientia experimentali: Chapters I–III*

It is important to give full attention to the first two chapters of Bacon's treatise on a science of experience in *Opus maius*, Part 6. These two chapters serve as an introduction to the three main prerogatives of a science of experience. They also serve to give context, structure, and finality to the three prerogatives. Bacon's sources in these chapters indicate that he is not a modern post-Cartesian scientist. He is very much a Medieval scientist who interprets Ancient and Medieval Islamic books of science, namely, Aristotle and his commentators, Seneca, Ptolemy, Ibn al-Haytham, Abu Ma'shar, and two works by two distinct Pseudo-Ptolemies, as well as many other writers.

Bacon begins by stating that having set out the foundations of Latin learning in languages, mathematics, and *perspectiva*, he will now examine the basis of a *scientia experimentalis*. Bacon is concerned with the epistemology of knowledge. And so, he states that there are two methods to gain knowledge: logical argument and experience. Logical argument infers a conclusion, and by necessity, we agree to this conclusion. But in matters pertaining to the heavens and the Earth, logic alone does not verify matters or remove doubt so that the mind rests in the vision of truth, unless that truth is gained by the method of experience. Given the central role of logic in Medieval university teaching and the absence of detailed work in natural science, many thought that logical argument alone was sufficient for empirical knowledge. Many have knowledge about things they have not seen. But as in the case of one who has never witnessed combustion, one who has only heard about fire may not avoid it unless he experiences the heat of the fire.

The need for experience is not limited to directly observable phenomena in the heavens and on Earth. Experience also has a role "in regard to that which has the strongest proof," namely mathematics. And bear in mind that in his treatment of mathematics in *Opus maius*, Part 4, Bacon "reduces" logic to mathematics. In *Opus maius* VI, Bridges' text gives just one sentence to Euclid's first proposition on the equilateral triangle:

> He who [mentally] has the strongest proof of the equilateral triangle without experience will never in his mind adhere to the conclusion, nor will he care but will neglect the proof until the experience of the equilateral triangle is shown to him by means of the intersection of two circles, from either intersection of which two lines may be drawn to the extremities of the given line. But then the man accepts the proof with all ease.[35]

Here, we are dealing with the teaching of geometry, something that gives expression to Bacon's teaching career. Bacon is concerned with pedagogy in the teaching of mathematics.

In the *Tractatus de experientia,* Bacon expands this one sentence into a long paragraph. It reads:

> For in mathematics, one finds the strongest proof and especially in geometry which is the foundation of all of mathematics. And the first among all the branches of mathematics is the stronger because through it the other branches gain strength because the conclusion of geometry is the principle or starting point in the other branches. If this proof [*demonstratio]* is made without experience, all lines from the center to the circumference are equal. But the first or second side of a formed triangle is led out from the center to the circumference like that line. Therefore, each side [of the triangle] is equal to the given line but if one and the same are equal, they are equal to themselves. On account of this, each side is not only equal to the given line but are equal to each other. And so the triangle has equal angles. Whoever receives this proof without experience [of the drawing of the triangle and circles] their mind will never adhere to this conclusion, nor will they care but will neglect it until an experience [an example] is given to them by means of two circles drawn to the extremities of the given line but then one will receive the conclusion with ease and will not ever again doubt it.[36]

Bacon then turns to Aristotle's account of demonstrative argument in the *Metaphysics, Posterior Analytics,* and *Nicomachean Ethics, VI.* Bacon's reading of Aristotle presents some interesting issues. The text reads:

> Therefore, Aristotle's statement that demonstration is a syllogism causing us to know [*Posterior* Analytics, 73a20-73a27] is to be understood provided that experience is attached and is not to be thought of as bare logical proof. His statement also in *Metaphysics* 1 that those who understand the reason and the cause are wiser than those who have experiential knowledge [*de expertis]* is spoken of experts who only know the bare truth/fact [that something is the case, *quia*] without a knowledge of the cause [*propter quid*]. But I am speaking here of the one who knows the reason and the cause through experience [*per experientiam*].[37]

This last sentence is important. After Grosseteste's commentary on the *Posterior Analytics*, all accepted that universals, the principles in scientific argument, were derived from repeated experiences and memories. The reason and cause, however, were then reached by direct inference or a syllogism causing us to know. And so, Aristotle does not need some additional experience to verify that proof.[38] But here, Bacon is talking about experiences which will verify the rational conclusions of arguments. Bacon continues: "These men [of experience] are perfect in their wisdom, as Aristotle states in the sixth book of the *Ethics*, whose simple statements must be believed/accepted as though they were proofs [demonstrations], as he states in the same place."[39] In the *Tractatus,* Bacon adds the following:

> As he himself determines in the book just mentioned. For as he himself states, [these mature experts] already have by [long] experience the knowledge by which they can provide the causes and proofs. Therefore, he who without doubt wishes to rejoice in the truth of things, ought to have time for [having] experiences. And if argument [alone, unaided by experience] is not adequate for the verification of truth even less so is mere authority, for it is weaker than argument or reason and the one who receives the power of things through reason receives a stable foundation. Therefore, this science [experimental] wishes to teach that one ought not to consider mere argument or authority unless it has some strong or light experience [accompanying] it.[40]

One thing is now clear: in a fundamental sense, Bacon is an Empiricist (in a realistic, although not a modern subjectivist, sense of the term). Or better, he requires, for knowledge, the combination of logical, mathematical proof, and experiential [empirical] verification. This experiential verification is something that Bacon shares with other writers on medicine such as Peter of Spain. The power of reason is retained: argument knows the power of things through reason and thus finds a firm basis. Therefore, this science [science of experience] wishes to teach that [bare] argument or authority alone is not to be entertained unless it is accompanied by strong or weak experience. For many authors write and the common scholar holds those things they imagine without experience, which are not merely doubtful but are entirely false.

It is clear from what follows that Bacon knew full well that many books on natural philosophy contained much that was exaggerated, legendary, and false.[41] His argument is that direct phenomenal experience is the only answer to such false claims.

The first example he gives of false empirical claims is the claim that diamonds can only be cut with the use of a goat's blood. Philosophers and theologians following Varro hold to this teaching. "And many wise persons and experts have attempted this, and do not realise that the common teachers assert doubts or falsehoods on this matter." Bacon answers that diamonds can only be cut with fragments of such gemstones. And he adds:

> For I have proved this by means of experience before my own eyes and this is necessary because gems cannot be cut without the fragments of gems. And uneducated persons who devote their time to experiences of gems-stones know this. [42]

Here, we note Bacon making a contrast between mere book-learning and learning based on practical experience. The next example has to do with false claims in medical books about the castors – the testicles of the beaver.

The example concerning the freezing of hot and cold water is very important. Here, Bacon is correcting a clear mistake in the Latin translations of this example in Book One of the *Meteoreologies* of Aristotle.[43] Translators and commentators claimed that Aristotle holds that hot water placed in a vessel freezes more quickly than cold water:

> But it is clear to anyone of experience that cold water freezes more quickly. People attribute this claim to Aristotle. But he certainly does not make this statement. But he does make one like it by which they are deceived, that is, that if cold water and hot water are placed on a cold place like ice, the hot water freezes more quickly, and this is true. Therefore, all things must be certified by the method of experience.[44]

But what then are the dimensions and limits of experience for Bacon? Bridges' text immediately states that "experience is two-fold." In the *Tractatus de experientia,* Bacon sets out more information before dividing experience into two kinds. He states:

> Similarly, famous Natural Philosophers and experts in *Perspectiva* believe that a mirror placed in water in a vase opposite the rays of the sun produce an appearance of the stars or the fixed stars or the spring

> sun joined to them, which is not far elongated from the area of the sun. Because such persons fail to experience things, they are mistaken as has been previously explained. And so, two idola (images) appear towards the rays of the moon, one greater and the other less whose speed no one can reckon. Therefore, because in these and in an infinite number of examples, people believe common speeches and fantastical arguments. Thus, they cannot be [regarded] as wise. And what is extraordinary is that in the text of Aristotle and his interpreters, many falsehoods are found which are serious in the bad translations on account of deficiency [on the part of translators] in experiences in the sciences and in [the observation] of [natural] things. Through these [experiential] ways this experiential speculative science teaches us [not to trust] in bare arguments or authorities without experience. Therefore, one ought to consider that all things should be verified by the truth of experience. And so, those who neglect experience can in no way understand. Because Sacred Scripture states he who is not a person of experience knows such things as if he states nothing with certitude.[45]

Bacon then introduces the next paragraph on the division of experience as twofold.

For Bacon, there are two distinct kinds of experiential knowledge. The first kind is through the external senses and in terms of Bacon's multiplication of species and consequent processing in the brain. This gives rise to a kind of rationality in the cogitative sense, which is important for rational knowledge. This kind of knowledge is human and philosophical. But even this latter knowledge is limited in its understanding of corporeal things. The second kind of knowledge is internal, illumined knowledge that is open to a revelation of new truths. These internal illuminations are internal intellectual knowledge, including an awareness of moral virtue, and do not depend on direct sense-knowledge alone. This illuminative knowledge is based on Pseudo-Ptolemy's *Centiloquium*.[46] For Bacon, quite independent of any special religious revelation, all knowledge, sacred and secular, has been revealed in and to the human minds of all wise persons. Bacon cites Pseudo-Ptolemy as follows: "There are two ways to arrive at a knowledge of things. One is through the experience of philosophy; the other is through divine inspiration, which is far better."[47] The text of the *Tractatus de experientia* adds the following: "And the one who has both of these ways to knowledge is perfect in wisdom as Ptolemy [Pseudo] and his Hali his Commentator [=Ahmed Ibn Yusur] declare."[48]

Following on this, Bacon claims that there are seven stages of this interior experience. Essentially, he connects scientific experience with moral truth and internal religious experiences. The goal here is to relate a science of experience to moral philosophy in the broad sense used by Bacon.[49]

In fact, for Bacon, moral teachings are spread among the sciences. He continues in the *Tractatus de experientia*: "There as briefly as I can I run through the highest roots of this science among the other sciences, and I show how more useful it is to theology [more] than any of the other speculative sciences."[50]

There are seven stages of interior knowledge: the first stage is the objectivity gained through scientific knowledge. The second stage is a knowledge of the moral virtues. Evil and ignorance can destroy scientific objectivity, a teaching which Bacon finds in Aristotle, Augustine, and Al-Ghazali. The ultimate model here is Socrates as honoured by Augustine in the book *De civitate dei.* The *Tractatus de experientia* adds the following: "And this [teaching about virtue and scientific knowledge] is clear from the ways [spiritual exercises] of the philosophers and their teachings as I will explain more abundantly in [the treatise] on morals."[51] Further, there are many parrot-like teachers who are blind to the truth. "Virtue, therefore, clarifies the mind so that a person understands more easily not only moral but also scientific truths."[52] The great example of this combination of moral and scientific knowledge is Bacon's disciple, the youth John.

The third stage of interior knowledge consists of the gifts of the Holy Spirit. The fourth is the beatitudes. The fifth is found in the spiritual senses. The sixth consists of spiritual fruits such as the peace of God beyond all understanding. The seventh is found in mystical raptures.

What follows is important with reference to the relation of moral and interior experience to knowledge of the sciences:

> And he who has been diligently exercised in these experiences or in several of them is able to ascertain for himself not only concerning spiritual matters but also in the human [and philosophical] sciences. [*Tractatus:* And each one ought to embrace these spiritual [and moral] experiences, not only through a spiritual life but also through the certification of wisdom not only in divine but also in human matters. Experience, therefore, is necessary for us; bare argument [without experience] is not sufficient.][53] Therefore, since all parts of speculative philosophy proceed by arguments, which result from topics based on authority or other topics based on argument alone, besides this which I now examine, it is necessary for us to have this kind of science which is called 'experimental. I wish to explain it not only as it is useful for philosophy but also for the knowledge of God, and for the rulership of the whole world, just as in the preceding parts I showed the relationship of languages and the sciences to the divine wisdom that guides all things.[54]

It is at this point that Bacon connects experimental science as the closest science to morals and claims that it is more essential than the other speculative sciences to theology. Bacon's reasons for this claim become very evident from what follows in both versions of his treatment of *scientia experimentalis*:

> Since this experimental science is totally unknown to the rank and file of the generality of students [of wisdom] I am therefore unable to convince people of its utility unless at the same time I reveal its power and its properties. This science alone knows how to experience perfectly what can be done by nature, what by the working of art, [*Tractatus:* what by human fraud, what by the operation of true spirits and what by the operation of fraudulent spirits], what the latter dream of and desire by songs, conjurations, invocations, deprecations, and sacrifices that belong to magic and [*Tractatus:* what truth is in them] so that all falsehood may be removed and the truth alone of art and nature be retained. This [experimental] science alone teaches us how to consider all the insane acts [and teachings] of magicians so that they may not be confirmed but be avoided, just as logic examines sophistical arguments so that they may be avoided.[55]

Experimental science is therefore a practical scientific method analogous to the logical method in argument. Where logic identifies sophistry in argument, experimental science will expose false empirical claims, legendary book-knowledge, false historical claims, and especially the fraudulent claims of magicians.

This reference is expanded in the *Opus tertium* (ed. Duhem), *Opus terium* (ed. Little), *De secretis operibus artis et naturae et de nulllitate magiae,* and his introduction to the *Secretum secretorum.* The Duhem and Little texts are similar, but what they add to the *Opus maius* is important. Bacon states that both theologians and philosophers forbid the invocation of demons. Only insane and malevolent magicians do so. And after demons are excluded, human fraud perpetrated by tricksters, subtle instruments, sleights of hand, songs, astrological characters, and other means should be excluded. These magicians, who seduce innocent people, do not understand that demons work through these means. This magic is not the work of art and nature. The Duhem and Little texts add:

> And all these books should be reviewed and condemned, such as the *Liber de Morte Animae, the Liber Fantasmatum, the Liber de Officiis et Potestatibus Spirituum,* and *the Libri de Sigillis Salomonis* and *Libri de Arte Notoria* and all such books which invoke demons or which proceed by means of fraud and vanity, but not by the methods of art and nature.[56]

The same list of magical books is given in the *Epistola de secretis operibus artis et naturae et de nullitate magiae.* To these lists, the introduction to the *Secretum secretorum* adds the following: *Liber Adae, liber Moysi, libri Salomonis et libri Aristotelis.* The remarks that follow the list in *Epistola de secretis operibus* are quite significant:

> It must be considered, however, that there are many books which are believed to be among the magical books but which [actually] contain

> the dignity of wisdom. Which ones are suspect, and which are not, will be demonstrated by the experience of any wise person. For if anything in any of them is found to be a work of nature or art, it should be accepted. If not, leave it as suspect. And since it is undignified and illicit for a wise person to practice magic, thus the latter is superfluous, and it is not necessary.[57]

In what follows, Bacon cites the authority of Isaac's *De Febribus*, [Pseudo] Aristotle's *Secretum secretorum,* Aristotle's *De metheora* III, and Book 10 of the *Nicomachean Ethics* that the rational human mind has the strength of virtue aided by divine influence to overcome being impeded in its knowledge and action, assuming the first proposition of the *Book of Causes* "that the primary cause has more influence on what is caused than does the secondary cause."

In the *Epistola de secretis operibus,* Bacon reinforces these philosophical positions with claims about future knowledge based on the *Centiloquium* and with special reference to the wonderful works of science, such as the example of the magnet, an issue close to the interests of his hero, the great experimenter, Magister Peter Peregrinus of Maricourt.

> And if it were not known to the world that a magnet attracts iron, this phenomenon would be regarded as a great miracle. But the experience of the wise man discovered this and besides this many works which the common students do not know.[58]

Magicians and others think this is mysterious and utter songs and sing over it, thinking that the power of the songs and signs draws the iron to the magnet. In truth, the natural power of the magnet attracts iron. "This is how it is, and I have proved it." And so, there are two important test-cases of the power of art and nature which for Bacon are associated with the great experimenter, Magister Peter, namely, a burning mirror, a scientific understanding of the magnet, and a new account of the construction of the astrolabe.

What then is the context for Bacon's attempt to save an art and science of nature and distinguish it from the arts of magic? It is clear from the above that he, and perhaps also Magister Peter, was directly trying to save the theory and practice of science in opposition to some theologians and Canon Lawyers who, in his view, condemned the good and the bad. Being ignorant of both magic and the sciences, they uttered general condemnations. And if one is to condemn books of magic, one must know what they contain, and which books are actually books of magic and not of science. Hence, for Bacon, there was a need to make a study of both books of magic and books of science in order to carefully preserve books which contain a true art and science of nature and to set aside books of magic.[59]

Bacon then turns to the first prerogative of experimental science, namely, the verification of the conclusions of the natural philosophers and the experts on *Perspectiva*. And the central example he picks is the study of the rainbow, halo, and related impressions as these have been presented in the *De methe-ora* of Aristotle, his thirteenth-century interpreters, and the texts of Seneca, Avicenna and Averroes.

1.6 The Structure and Content of Bacon's *De scientia experimentali* as Found in *Opus maius*, Part 6, (ed. Bridges) and the *Tractatus de experientia in communi* (ed. Hackett)

It will be very helpful to review and compare these two versions of *De scientia experimentali*. Here, we are perhaps dealing with two separate drafts of Bacon's text. The final text that Bacon sent to the Pope is not extant. There are some lengthy passages in the *Tractatus* that are not present in Bridges' edition. This is particularly the case for the first three chapters. On the other hand, for Chapters 4 to 6 on the rainbow, one has virtually the same text. However, Chapters 7 to 12 are missing from the *Tractatus*. After Chapter 6, the writer goes immediately to the second prerogative. And whereas in Bridges' text, the division into three prerogatives is signalled at the end of Chapter 2, this is not so in the *Tractatus*. In the latter, the account of the rainbow is presented as a great example of how some of Bacon's contemporaries have misunderstood Book 3 of Aristotle's *Meteorologies*. Much of the additional material in the introduction has to do with Bacon's criticism of the Latin text of this latter work and with contemporary interpretations of the text.

It is clear from Bridges' text that Bacon intended the work to have three prerogatives. In Chapter 2, Bacon simply states:

> This science has three leading prerogatives with respect to the other sciences. The first is that it investigates all the noble conclusions of the other sciences by means of experience. For the other sciences know how to discover their principles by means of experiences (*experimenta),* but their conclusions are reached by means of the discovered principles. But if they should have a particular and complete experience of their own conclusions, they must have it with the help of this noble science. For it is true that mathematics has universal experiences of its conclusions in its (geometrical) figures and numbers, which also are applied to all sciences and to this experience, because no science can be known without mathematics. But if we give our attention to particular and complete experiences and are generally verified in their proper discipline, it is necessary to proceed by means of this science which is antonomastically called experiential (*experimentalis).* And I provide an example of the rainbow and impressions connected with it such as the circle around the sun and the stars, and also the streak also lying at the side of the sun or a star which is apparent to the eye in a straight line, and is called by Aristotle in the third book of the *Meteorologies* a perpendicular, but by

Seneca a streak, and the circle is a corona, phenomena which frequently have the colour of the rainbow.[60]

Experts in natural philosophy discuss these phenomena, and the teachers of *Perspectiva* examine the mode of vision in viewing these phenomena. But neither Aristotle, Seneca, nor Avicenna has given an adequate account of these matters. In the *Tractatus*, one finds a page and a half of additional comment on the inadequacy of the natural philosophers:

> But their speeches are useless, and conclusions are reached through arguments without sufficient reference to experiences. And so, they generate no certitude in the hearts of the audience and not only are doubts found among the teachers but also falsehoods. For in the translation of the books of the *Meteorologies* spread among the Latins up to now it is said that a lunar rainbow does not occur except twice in fifty years. And the greatest natural philosopher and expert on *perspectiva* whom I saw, wished to save this truth and provide a cause for this while I was already his student from my youth. But he was seduced by arguments and false translation, since experience teaches that in every full moon when the moon shines without any blockage of clouds or vapors, and the proper matter is generated opposite the moon, an occurrence of the rainbow will be generated, just as is the case with the halo and the perpendicular line which is thought to be a third coloured impression. And although the text of Aristotle proposes that this perpendicular is a distinct impression and Seneca likewise in this affirms the same thing about these impressions, the man of experience however knows that what is called a perpendicular by Aristotle and a streak by Seneca is not a new and distinct impression from the halo as afterwards the man of experience will show, because he sets aside authorities and bare arguments by using [scientific] instruments. And when all crude natural philosophers barely exercised in [scientific] experiences say that the colours of the rainbow and halo come about due to the diversity in the thickness and rarity of the cloud according to what the bad translation of Aristotle teaches. But neither Aristotle nor Seneca nor Avicenna give a good account of these matters. Indeed, much useless information is contained in their books and certain falsehoods are included as will be clear from what has been said.[61]

Bacon then, in the two texts, states that the person of experience must provide an adequate description of the various rainbow phenomena in the sky and by means of crystalline stones. He then proceeds to give his own account of the rainbow and related phenomena. What will become crucial in Bacon is the combination of mathematics, the careful correlation of individual phenomena, and the skilful use of instruments such as the astrolabe in observing the sky. In this way, the rainbow and related phenomena will provide the main example for the efficacy of the first prerogative of experimental science.

When we turn to the Bridges edition of *De scientia experimentali* and the Hackett edition of the *Tractatus de experientia in communi* concerning the second and third prerogatives, these prerogatives are presented in quite different terms in both versions of *De scientia experimentali.* Immediately after Chapter 12, Bridges' text has the following heading: "*Capitulum de secunda prerogative scientia experimentalis.*" The opening paragraph states that experimental science, while using the terminology of other sciences, can discover truths that these sciences themselves have not found. And there are many examples of this. In this matter, it is not enough for the inexpert to ask for bare reasons. In this matter, experience is crucial. Methodologically, one believes the expert, then one seeks the experience, and then one invokes the reason. Bacon then provides examples of this methodology: the armillary sphere to map the heavens, an account of the prolongation of life in medicine, and the development of new metals in alchemy.

Bridges' text opens the discussion of the third prerogative as follows: "*Capitulum de tercia prerogativa vel dignitate artis experimentalis.*" This prerogative deals not with the other sciences but with how experience itself investigates the secrets of nature. And it has two parts. The first has to do with knowledge of the future, the past, and the present. The second has to do with forming astrological judgements that are better than "the ordinary astronomy dealing with judgements." However, Bridges' text is very brief. He then turns to various technologies that can be useful in warfare and military life and concludes with a brief account of the utility of a science of experience for the Church and Commonwealth.

In the *Tractatus*, however, Bacon introduces the second prerogative as follows:

> Apart from the first prerogative this science has other prerogatives and dignities in reference to the other sciences which those sciences cannot complete without the benefit of this (experimental) science. Although they are in the terminology of these other sciences, these sciences are not able to find them or to produce them completely. Such sciences are (1) the science of the prolongation of life, (2) the science or knowledge of the future, [based on the *Centiloquium*], (3) the science of the transmutation of metals and (4) similar sciences.[62]

On completion of a short summary of the prolongation of life, Bacon turns to a knowledge of the future based on Ptolemy's *Quadripartitum, De dispositione spherae, Centiloquium.* These texts provide the means to improve on the ordinary judgements of astronomy. This text corresponds to the material found in the introduction to the *Secretum secretorum* and elsewhere. Bacon then provides a very brief outline of teaching on alchemy. What follows this account of alchemy comes as a surprise. He writes:

> The Third Noble Dignity [or Prerogative] of this practical science determines its speculative part through works which are proper to this science, which are not used by the other sciences, nor do they aspire to those works. And I will give an example of which I am certain. For it teaches about perpetual baths and perpetual lights and living wheels.[63]

Bacon then talks about the construction of instruments and more efficient technologies of war. All this account is the same as found in Bridges' text immediately after the brief identification of the Third Prerogative with knowledge of the future by means of experiential judicial astrology. And so, based on the *Tractatus,* one would have to argue that the third prerogative of experimental science has to do with new technologies of public welfare and of warfare. And that the account of the knowledge of the future by experiential astrology is just one among four examples of the Second Prerogative of the use of experimental sciences within the sciences. In other words, it seeks to improve and make more precise the judgements of astrology.

Here, it is useful to turn to Bacon's division of astrology/astronomy in the *Communia mathematica.* This material should be coordinated with the treatise on astrology that is printed in the final section of *Opus maius,* Part 4, and which H. Darrel Rutkin, with good reason, identifies as part of the *Opus minus.*[64] First, in his division of mathematics, Bacon notes that:

> All matters which are under consideration by Astrology and Astronomy are certified by the [mathematical] methods of geometry and arithmetic as Ptolemy holds in the first chapter of the *Almagest* and all things in music are made clear by the methods of numbers and figures as is clear to experts in music.[65]

Thus, geometry and arithmetic are prerequisites for astrology and its practice, more so than music.

In the fifth chapter, Bacon provides a division of astrology/astronomy. It is divided into speculative astrology and its practical applications, which is named astronomy. Speculative astrology teaches two things: the verification of the magnitude of the Earth and its shape, along with its habitable and non-habitable parts. The other part is the verification of the heavenly bodies in terms of motion, place, time, number, magnitude, figure, altitude, and degree of density. And the causes of these bodies are presented by Ptolemy in the *Almagest* with additional help from Geber, Thebit, and from the books of Alfraganus, Marianus, along with other teachers of proven repute by means of a narrative account.[66] The practical extension of this astrology is called astronomy and is concerned with gaining certitude about the place of the planets and fixed stars and orders the arithmetician and geometer to construct the necessary tables and instruments. This involves much work and great cost. This practical astronomy is not only concerned with:

> The heavenly bodies, but with all things that arise above the horizon such as the impressions already mentioned. For it falls to the mathematical astronomer to verify such matters in so far as they come to bear in the use and experiential knowledge of the future, the present and the past, so that bad understandings can be impeded and good knowledge can be promoted, both for the commonwealth and persons.[67]

Practical astronomy, unlike speculative astronomy, is not a matter of general narrative speech. It has to do with the applications of astronomical laws. Thus, practical astronomy, more so than general astrology, devolves to matters of practical application. As Isidore states in his *Etymologies,* one part of astronomy is natural, while another part is "superstitious," explaining those things which incline themselves to judgements. And, so, less knowledgeable people deviate into the "superstitious" part. But this will be treated with greater certainty when I examine these sciences. It is clear then that, like Augustine and Isidore, Bacon aims to make a strict distinction between the art and science of nature and an astronomy that is superstitious and magical.

To conclude, Roger Bacon links the concerns of Aristotle in the *Meteorologies* with the unified project of Ptolemy in the *Almagest* and the *Quadripartitum.* But he also takes over the general work of Ptolemy on geography, time-calculation, history, and optics. Even those important Pseudo-Ptolemaic works cited above, especially the *De dispositione spherae,* the *Centiloquium,* and the commentary by Haly [=Ahmed Ibn Yūsuf], are linked to Ptolemy. We can acknowledge then that Roger Bacon is very much a child of his own times and is not a post-Cartesian scientist. Nevertheless, his skilful mixing of Aristotle, Ptolemy, and Pseudo-Ptolemy enabled him to emphasise the experiential dimensions of knowledge. And while he duly praises the role of inference in knowledge, he argued that bare inference without confirmation by experience would not work in science.

This chapter has examined the technical meaning of the term *scientia experimentalis*, Bacon's own acknowledgement of the limits of his own knowledge of *Perspectiva*, and the identification of *Magister Peter* as the true experimenter. It examines Chapters 1–3 of *Opus maius*, Part 6, which serves as an introduction to the three prerogatives and emphasises the importance of *scientia experimentalis* as a method to distinguish an art and science of nature from the arts of magic. It also compares the two extant versions of *Opus maius*, Part 6, with regard to the three prerogatives of experimental science. It concludes by presenting treatises on astronomy/astrology that are closely linked to *scientia experimentalis.*

Notes

1 For a review of these sources, see Hackett, "Ego Expertus Sum," 145–174 (esp. 146–147).
2 Lindberg, "Roger Bacon on Light," 243–275.

3 Hackett, "Roger Bacon on Scientia Experimentalis," 277–315.
4 Panti, "The Theological Use of Science," 143–163.
5 Allen, *Roger Bacon and the Incorruptible Human.*
6 Kedar and Hon, "Roger Bacon and his System of Laws of Nature," 719–745.
7 See Hackett, "Roger Bacon's Various Texts on Perspectiva," 81–104. See Paris, BN Lat. 7434.
8 Bacon, *Secretum secretorum.* This work was most likely completed at Oxford ca. 1280, but the introduction and notes indicate that Bacon already had completed some of the material in the 1260s. For more on Bacon and magic, see Nicolas Weill-Parot's chapter in this volume.
9 Even though Bacon cites *De dispositione spherae,* the content referred to is from Pseudo-Ptolemy, *Centiloquium,* and the *Expositio* by Ahmed Ibn Yusuf. See unpublished edition of the *Centiloquium* (Iam Scripsi version) edited by Richard Lemay, Irvin Department of Rare Books and Special Collections, Hollings Library, University of South Carolina. That knowledge of the future is twofold joined with the reference to Arathus would seem to be from the *Iam scripsi* version of *Centiloquium, Verbum 1 and 4.* Still, the *Aratus* reference in the *Iam scripsi* version and the *Mundanorum* version are similar.
10 Pseudo-Ptolemy, *Centiloquium*, 85. ALNAEZIC significant siccas vapores. Cum ergo fuerint omnes in una parte significabunt ventos accirentes in eadem parte. 86 [Exposito}: aristoteles docuit nos in Libro quem composuit *De hiis que superius accidunt* [*Metereologica*] quod basi sicci quando ascentunt usque ad cierculum igneum faciunt istas scintillationes (871) ideo sigtnificant siccitatem et minutionem aquarum. But it would seem from Bacon's reference to *Assub* that he is using the *Iam scripsi* Version and not M. See *Iam scripsi*, [Verbum 100, Expositio], ed. Lemay, F44, Iam patefecit Aristoteles in *libro de operibus altis* quod vapores aridi quociens perveniunt ad etherem fiunt *assub* et sunt almanezic.
11 Aristotle, *Meteorologies*, 15–18. Bacon fuses the empiricism of Ptolemy, Aristotle, and Pseudo-Ptolemy, *De dispositione spherae*, and Pseudo-Ptolemy, *Centiloquium.* On the importance of the *Meteorologies* for Aristotle's science, see Falcon, *Aristotle and the Science of Nature.*
12 Aristotle, *Metereologies*, Bk. III.
13 Pseudo Ptolemy, *Centiloquium*, [Verbum 1, M]. Anima naturata iudicat super secundis a stellis (58), et erit eius certificatio plus quam multarum (59) secundum ipsa sstraa iudicantium. Dixit Ptholomeus: Anima naturata etc. Anima que ex natura dat iudicia iudicabitur super secundas stellarum eiusque iudicium erit melius quam illius qui iudicat per illas (60) stellas. [Expositio] Iam docuit quomodo anima habet hoc ex natura. Nunc docet Ptholomeus quod secunde stellarum sunt ex opere quod ipse faciunt sub circulo Lune, ut sunt scintillationes et stelle cum caudis, scilicet comete, et circulus qui apparet quandoque circa Lunam, quandoque circa Solem, et alia multa que prope circulum (61) inferius sunt. Et hec omnia (62) nominavit Arathis [...;] The brevity of Bacon's citation might indicate that he is using the *Iam scripsi* version and not M. See I, 3. Dixit Ptolomeus. Anima que ex natura dat iudicia iudicabit super secundas stellarum eritque eius iudicium melius quam illius quid iudicabit per ipsas stellas. And since the reference to Arathus follows immediately, the reference to Arathus in n. 9 above most likely also comes from the *Iam scripsi* version. For more on Bacon and astronomy, see H. Darrel Rutkin's chapter in this volume.
14 Pseudo-Ptolemy, *Centiloquium*, 4: [M, Verbum 1, Expositio]. Hoc quod dixit Ptholomeus: "astrologus non debet dicere rem specialiter sed universaliter" ideo dixit quia hec scientia non est nisi secundum probationem et opinionem. See also version I, [Verbum 1, Expositio], 1: Hoc quod dixit Ptholomeus "astrologus non

debet dicere rfem specialiter sed universaliter ideo dixit quia hec scientia non est nisi secundum probationem et opinionem.

15 Ptolomey, *Tetrabiblos*, 2–35.

16 Bacon, *Opus maius*, vol. I, 243–245.

17 Bacon, *Opus maius*, vol. I, 245–246. Ex quibus verbis et consimilibus manifestum est, quod de intentione Ptolomei non est quod astrologus det in particulari certum iudicium et sufficiens in singulis...ut dicit Messehalac ponens exemplum de magnete. See Birkenmajer, "Avicenna's Vorrede zum "Lliber Sufficientiae" and Bacon, *Revue néo-scholastique de philosophie, 308–320. See 319:* Deinde consequenter scienciam naturalem ordinavi, ubi evenit mihi in pluribus rebus imitacio edicionis Aristotelis [....] Deinde consequenter adiunxi geometriam, abreviando librum *Elementorum* Euclidis [....] Deinde consimilem apposui abreviacionem libri *Almagesti* in astrologia, conntinentem cum abreviacione intellectualem ac explanacionem.

18 See Toomer, *Ptolemy's Almagest*, 35–47.

19 See Feke, *Ptolemy's Philosophy*, 10–35. But see 19: "Although Ptolemy's definitions of the sciences are not Aristotle's, they are still Aristotelian." Most important for Bacon is Ptolemy's teaching that physics and theology are matters of conjecture aided by mathematics. This changes Aristotle considerably.

20 Bacon, *Communia mathematica*, 8: [...] non parum valet mathematica ad reliquorum duorum modorum sciencie comprehensionem, ut docet Ptholomeus in primo capitulo *Almagesti*.

21 Bacon, *Communia mathematica*, 8.

22 Pseudo-Ptolemy, *Centiloquium*, 3–4. See I, [Verbum primum, Expositio]. quod dixit "hec iudicia esse inter necessarium et possibile" hoc est apud eum qui rerum naturas et stellarum opus considerat. Quod dixit Ptolomeus "ex meliori parte" dixit secundum vim anime qua rationales sumus. Et in illa vi est natura qua ipse cognoscit quid contingere bebeat. Et quia habemus contra vim iram et cupiditatem et quia oportet nos esse sollicitos circa mundana, aufertur a nobis futurorum cognitio. Sed si hec vis a mandanis omnibus separetur possemus futura porenoscere. Sicut videmus heremitarum quamplures qui futura predicunt, et epilenticorum quidam dum epilelmpsia torquentur, quia tunc corpporeis sensibus non utuntgur sed sola vi anima quandoque futura prenuntiant. This latter sentence is used by Bacon in the context of the third prerogative of experimental science. Again, it appears that Bacon favours the *Iam scripsi* version.

23 Bacon, *Opus maius*, vol. II, p. 201. Et ideo nullus sermo in his potest certificare; totum enim dependet ab experientia. Et propter hoc non reputo me attigisse hic plenam veritatem; *QUIA NONDUM EXPERTUS SUM* omnia quae sunt hic necessaria, et quia in hoc opere procedo via persuasionis et ostensionis quae oporteat requiri in studio sapientiae, et non per viam componendi scripta de ea. Et ideo non pertinet ad me hac hora dare certificationem impossibilem, sed sub forma persuadendi de studio sapientiae pertransire. On Bacon's use of the word *Expertus,* see Bacon, *Compendium studii philosophiae*. De qua est causa perversitas translationis maxime in libris Aristotelis et scientiis eius [...] ut nullus mortalis possit alilquid Dignum de eis intelligere veraciter, *sicut Ego Expertus Sum omnino*.

24 Bacon, *Opus minus* [forthcoming]. I thank Professor Maloney for the use of this translation.

25 Bacon, *Opus tertium*, 43. Sed praeter has scientias est una perfectior omnibus cui omnes famulantur, et quae omnes miro modo certificat: et haec vocatur scientia experimentalis, quae negligit argumenta, quoniam non certificant, quantumcunque sint fortia, nisi simul adsit experientia conclusioinis, ut ostendo in tractatu de ista scientia. See Bacon, *Opus tertium*, 86–87. The reference to the "experience

of the conclusion" must be distinguished from experience and memory as the source of the experimental universal in Aristotle.

26 Bacon, *Opus tertium*, 43–44. Naturales quidem in libro Metheororum Aristotelis, et perspectivi, negotiantur circa haec certficanda sed in vanum; quia sola experientia certificat hic, et non argumentum. Et ideo pono radices experimentiarum circa ista, *quas nullus Latinorum potest intelligere nisi unus, scilicet, Magister Petrus. Et sic est de aliis rebus naturalibus. Alia est dignitas huius scientiae, quod certificat veritates nobiles in terminis aliarum scientiarum, quae non possunt ibi probari, nec aliquo modo investigari; ut est prolongatio vitae humanae, ut multae aliae; nam haec veritas est in terminis medicinae, sed nunquam ars medicinae se extendit ad hanc veritatem, nec est aliquid de hoc in libris medicinae.* ed. Egel, 88–89.

27 Bacon, *Opus tertium*, 44. Tertia vero dignitas istius scientiae est, quod non recipit veritates in terminis aliarum scientiarum, sed tamen utitur eis sicut ancillis. Et haec pars huius scientiae extendit se ad cognitionem futurorum, et ad operationem mirabilium operum naturae et artis. ed. Egel, 90–91.

28 This is based on a combination of Ptolemaic and Pseudo-Ptolemaic texts. See note 2 above.

29 Bacon, *Opus tertium*, 45. Et Gratianus et multi minus bene dixerunt in hac parte, quia non omnia sunt magica quae ipsi docent reprobari, et reprobant falsa quae ignorabant. Sed homo, qui reprobat aliquid, debet scire eius condiciones, et sic falsum reprobare, ut veritas semper maneat illaesa. Egel, 90–91.

30 Bacon, *De speculis comburentibus*, 271–341. This skillful text suggests that Bacon may have been the geometric advisor to Magister Peter's experimental work on crafting a burning mirror.

31 See Hackett, "Roger Bacon on Military Technology," 123–135.

32 Magister Petrus Peregrinus de Maricout, *Epistolae de Magnete*. See Nagel, "Pietro Peregrino," 21–47. This study provides a careful comparison of Magister Peter and Roger Bacon's *scientia experimentalis*.

33 Bacon, *Opus tertium*, 46–47; Egel ed., 94–97.

34 As will become apparent below, Bacon was involved in a project to separate books on the art and science of nature from books on magic. This is not the place to provide a formal comparison between Bacon's texts studied in this chapter and the *Speculum astronomiae*. I will examine this matter elsewhere. However, these Bacon texts contain strong suggestions that Bacon worked with those who wrote this text, which makes a formal distinction between a science of nature and books on magic.

35 Bacon, *Opus maius*, vol. II, 168. Qui vero habet demonstrationem potissimam de triangulo aequilatero sine experientia nunquam adhaerebit animus conclusioni, nec curabit, sed negliget usquequo detur ei experientia per intersectionem duorum circulorum, a quorum alterutra sectione ducantur duae lineae ad extremitates lineae datae; sed tunc recipit homo conclusionem cum omni quiete.

36 Hackett, "A Working Edition of Roger Bacon," 273–274. (I will refer to this work as Hackett, ed., *Tractatus de experientia)*. Nam in mathematicis est demonstratio potissima et maxime in geometria quae est fundamentum omnium mathematicalium. Et prima inter omnes est fortior quia per illam ceterae sumunt vigorem, quoniam eius conclusio est principium probandi cetera. Si igitur fiat haec demonstratio sine experientia, omnes lineae ductae a centro ad circumferentiam sunt aequales. Sed utrumque latus constituti trianguli super lineam datam egreditur a centro ad circumferentiam sicut illa linea. Ergo utrumque latus est aequale lineae datae, sed quae unum et eidem sunt aequalia sibi invicem sunt aequalia, quapropter utrumque latus non solum est aequale lineae datae sed ad invicem sunt aequalia, et ideo triangulus erit equiangulus. Quicunque vero hanc demonstrationem potissimam habet sine experientia, adherebit animus conclusioni nec curabit,

sed negliget eam iis quoque datur ei experientia eius per intersectionem duorum circulorum a quorum alterutra sectione ducantur duae linee ad extremitates linee datae. Sed tunc recipiet homo conclusionem in quiete et non poterit ulterius dubitare.

37 Bacon *Opus maius*, vol. II, 168. Quod ergo dicit Aristoteles quod demonstratio syllogismus est faciens scire, intelligendum est si experientia comitetur, et non de nuda demonstratione. Quod etiam dicit primo *Metaphysicae*, quod habentes rationem et causam sunt sapientiores expertis, loquitur de expertis qui solum noscunt nudam veritatem sine causa. Sed hic loquor de experto, qui rationem et causam novit per experientiam.

38 On the importance of *experimenta* that verify formal proof, see Crisciani, "Experimentum, ricetta, raccolta," 223–242. The importance of medicine and alchemy for Bacon's epistemology of experience is emphasised by Meagan S. Allen in her book-length study mentioned above.

39 Bacon, *Opus maius*, vol. II, 168. Et hi sunt perfecti in sapientia, ut Aristoteles vult sexto Ethicorum, quorum sermonibus simplicibus tunc credendum est ac si afferrent demonstrationem, ut dicit ibidem.

40 Bacon, *Tractatus de experientia*, 276. Ut ipse in sexto libo nunc tacto determinat. Nam ut ait ipse, habent iam per experientiam unde possunt causas dare et demonstrationes afferre. Qui igitur vult sine dubitatione gaudere de veritatibus rerum, oportet quod experientiae sciat vacare. Etsi argumentum non sufficit ad certificationem veritatis longe minus auctoritas sufficit, quoniam debilior est ratione et argumentatione habet virtutem rerum per rationem recipiat stabile fundamentum. Haec igitur scientia vult docere quod non est considerandum argumento aut auctoritati nisi aliqua experientia fortis vel levis habeatur.

41 For a good example of Bacon's commitment to an experiential and evidence-based history, see *Opus maius*, Part 4, 304–305. Non numquam tamen plura reperiuntur scripta, quae ex rumore magis quam per experientiam habuerunt auctores. The observations of the Tartars by William of Rubruck are presented as one based on reliable historical scientific observation.

42 Bacon, *Tractatus de experientia*, 277. Et plures sapientes et experti tentaverunt hoc et non invenerunt quapropter vulgus falsum aut dubium asserit in hoc casu. Et tamen sine ullo sanguine potest frangi de facili. Hoc enim probavi per experientiam ante oculos meos et necesse est hoc quia gemmae non possunt sculpi nisi per fragmenta huius lapidis. Et hoc sciunt rustici qui dediti sunt experimentis gemmarum.

43 Aristotle, *Meteoreologies*, 280.

44 Bacon, *Opus maius,* vol. II, 169. Sed certum est quod aqua frigida citius congelatur experienti. Et imponunt hoc Aristoteli secundo Metereologicorum, sed pro certo non dicit hoc. Oportet ergo omnia certificari per viam experientiae.

45 Bacon, *Tractatus de experientia*, 280–281. Similiter famosi naturales et perspectivi credunt quod speculo posito in aqua in vase contra radios solis fit apparitio stellae cum sole vel stellae fixae vel veneris solem concomitantis quae non multum a solis vicinis elongatur, quia negligunt candelam experiri, decipiuntur ut prius expositum est. Et ad radios lunae similiter apparent duo idola, unum maius et reliquum minus cuius velocitatem nulla stella potest concomitari. Quia igitur in his et infinitis credimus sermonibus vulgaribus et fantasticis argumentis, non possumus esse sapientes. Et quod mirum est in textu Aristotelis et suorum expositorum, multa reperiuntur falsa quae tamen forte sunt de malis translationibus propter defectum experientiae in scientiis et in rebus. Per has igitur vias docet haec scientia experimentalis speculativa quod non est in argumentis aut auctoribus sine experientia. Considerandum oportet igitur certificari omnino per veritatem experientiae. Et ideo qui negligit experientias nihil potest scire propter quod Scriptura

querit Sacra: Quia non est expertus qualia scit quasi dicat nihil secundum certitudinem.

46 Pseudo-Ptolemy, *Centiloquium*, [Verbum 1, Expositio, *Iam Scripsi* version, 1.]. For a study of this internal knowledge, see Boudet, "Astrology Between Rational Science and Divine Inspiration," 49–75. See 67, (n.37) for Bacon's use of *verba* 5 and 8 of the *Iam scripsi* version in *Opus maius*, Part 4, 391. Again, it would seem from that and the examples from the introduction to the *Secretum secretorum* that Bacon favoured the *Iam scripsi* version.

47 Pseudo-Ptolemy, *Centiloquium*, [Verbum1, Expositio], 4.

48 Bacon, *Tractatus de experientia*, 283. Et qui habet utramque perfectus est in sapientia ubi ipse et Haly commentator eius declarant. See *Centiloquium*, [Verbum 1, Expositio], 4.

49 See Bacon, *Moral Philosophy*.

50 Bacon, *Tractatus de experientia*, 289. Quam brevissime igitur possum transcurram eius radices altissimas inter scientias et ostendam quomodo utilior est theologiae quam aliqua aliarum de speculativis scientiis. Just prior to this, he writes: Nobiliorum autem scientiarum et utiliorum post moralem, dico esse scientiam experimentalem.

51 Bacon, *Tractatus de experientia*, 284. Et sicut patet per viam philosophorum et sententias eorum et in moralibus uberius exponetur.

52 Bacon, *Opus maius*, vol. II, 171. Virtus ergo clarificat mentem ut non solum moralia sed etiam scientialia homo facilius comprehendat.

53 Bacon, *Opus maius*, vol. II, 171. Et qui in his experientiis vel in pluribus eorum est diligenter exercitatus, ipse potest certificare se et alios non solum de spiritualibus, sed omnibus scientiis humanis; *Tractatus de experientia*, 288. Et quilibet deberet ad has experientias anhelare, non solum per vitam spiritualem sed per certificationem sapientiae non solum divinae sed humanae. Experientia igitur necessaria est nobis et non sufficit argumentum.

54 Bacon, *Opus maius*, vol. II, 171–172. Et ideo cum omnes partes philosophiae speculativae procedant per argumenta, quae vel fiunt per locum ab auctoritate vel per caeteros locos argumentandi praeter hanc quam nunc investigo, necessaria est nobis scientia, quae experimentalis vocatur. Et volo eam explanare, non solum ut utilis est philosophiae, sed sapientiae Dei, et totius mundi regimini, sicut in pri oribus comparavi linguas et scientias ad eum finem, qui est divina sapientia qua omnia disponuntur.

55 Bacon, *Opus maius*, vol. II, 172. Et quia haec Scientia Experimentalis a vulgo studentium est penitus ignorata, ideo non possum persuadere de ejus utilitate, nisi simul ejus virtus et prorietas ostendantur. Haec ergo sola novit perfecte experiri quid potest fieri per naturam, quid per artis industriam, quid per fraudem, quid volunt et somniant carmina, conjurationes, invocationes, deprecationes, sacrificia, quae sunt magica, et quid fit in illis ut tollatur omnis falsitas et sola veritas artis teneatur. Haec sola docet considerare omnes insanias magicorum non ut confirmentur se ut vitentur, sicut logica considerat sophisticam artem. This text is repeated by Bacon in many other places.

56 Duhem, *Un Fragment Inédit*, 151–152; Little, *Part of the Opus tertium*, 148.

57 Bacon, *Epistola de secretis operibus artis et naturae*, 532. Considerandum est tamen quia multi libri reputantur inter magicos qui non sunt tales, sed continent sapientiae dignitatem; qui igitur sunt suspecti et qui non, experientia cuiuslibet sapientis docebit. Nam si quis in aliquo illorum opus naturae vel artis inveniat, illud accipiat; si non, relinquat velut suspectum.

58 Bacon, *Epistola de secretis operibus artis et naturae*, 537.

59 For a careful contextual study of Roger Bacon's understanding of magic when compared to Albert the Great and

Thomas Aquinas, see Weill-Parot, *Les "images astrologiques" au Moyen Âge*, 303–339; see Weill-Parot's lecture *on Bacon's condemnation of Magic* on the website of the *Roger Bacon Research Society.*

60 Bacon, *Opus maius*, vol. II, 172–173. Et haec scientia habet tres magnas praerogavitas respectu aliarum scientiarum. Una est quod omnium illarum conclusiones nobiles investigat per experientiam. Scientiae enim aliae sciunt sua principia invenire per experimenta, sed conclusiones per argumenta facta ex principiis inventis. Si vero debeant habere experientiam conclusionum suarum particularem et completam, tunc oportet quod habeant per adjutorium istius scientiae nobilis. Verum est enim quod mathematica habet experientias universales circa conclusiones suas in figurando et numerando, quae etiam applicantur ad omnes scientias et ad hanc experientiam, quia nulla scientia potest sciri sine mathematica. Sed si attendamus ad experientias particulares et completas et omnino in propria disciplina certificatas, necessarium est ire per considerationes istius scientiae, quae experimentalis auctoritate vocatur. Et pono exemplum in iride et ei annexis, cuiusmodi sunt circulus circa solem et stellas, virga quoque jacens a latere solis vel stellae, quae apparet visui in linea recta, et vocatur ab Aristotele tertio Metereologicorum... corona, quae pluries habent colores iridis.

61 Bacon, *Tractatus de experientia*, 292–294. Sed sermones eorum inutiles et per argumenta sine experientiis conclusioni sufficientibus. Et ideo nullam certitudinem generant in cordibus auditorum et non solum dubia inveniuntur apud doctos sed falsa. Nam in translatione libri Meteorologicorum pervulgata apud latinos usuge nunc, dicitur quod a radiis lunae non sit iris nisi bis in quinquaginta annis: et maximus naturalis et perspectivus quem vidi, voluit et hoc verum salvare et causam eius reddere dum eius auditor a iuventute fueram constitutus. Sed ipse argumentis seductus et in translatione falsa, quoniam experientia docet quod in omni plenilunio si luna luceat sine impedimento nubium et vaporum, et materia rorida sit generata in oppositione lunae, generabitur iris accio, sicut est de halo et linea perpendiculari quae figitur esse tertia impressio colorata. Sed licet textus Aristotelis pronunciet hanc perpendicularem tamque distinctam impressionem, et Seneca similiter in hoc de hiis impressionibus hoc idem affirmat, tamen experimentator scit quod id quod vocatur perpendicularis ab Aristotele et virga a Seneca non est impressio nova et distincta ab halo sicut postea exponit experimentator, quia suspendit auctoritates et argumenta per instrumenta, et cum omnes rudi naturales et parum exercitati in experientiis dicunt hos colores iridis et halo fieri penes diversitatem nubis in spissitudine et raritate secundum et quod mala translatio Aristotelis innuit. Sed nec Aristoteles nec Avicenna in suis "Naturalibus" huius rerum notitiam nobis dederunt, nec Seneca qui in hiis librum composuit specialem. Quoniam multa inutilia continentur in libris eorum et aliqua falsa inseruntur sicut ex nunc dicendis elucescat. Perhaps, Bacon is referring to the versions of the *Meterologica* translated by Gerard of Cremona. But in his *Opus maius,* Part 6 on experimental science, Bacon deals with the "nova" translation of William of Moerbecke. Bacon's account of the rainbow and halo offers a criticism of Moerbecke's translation.

62 Roger Bacon, *Tractatus de experientia*, 332–333. Habet praeterea haec scientia secundo loco alias praerogativas et dignitates aliarum scientiarum, scilicet, quantum ad opera sapientiae quae non possunt compleri sine beneficio istius scientiae quae licet sint in terminis aliarum scientiarum, tamen aliae scientiae non possunt eas procurare, nec completum perducere. Huiusmodi sunt prolongatio vitae, scientia futurorum, scientia transmutationis mineraliun et similes.

63 Bacon, *Tractatus de experientia*, 344. Tertiam dignitatem et nobilitatem huius scientiae practicae determinat eius speculativa per opera quae sunt huic scientiae propria, quibus non utuntur aliae scientiae, nec aspirant ad illa, sed ponam hic ali-

qua exempla de quibus certus sum. Nam docet facere balnea perpetua et luminaria perpetua et rotas vivas, id est, continue motas et hoc variis modis.

64 Bacon, *Opus maius*, vol. II, 376–404; See Rutkin, "An Idealized Astrological Courtier," 53–69. See also Rutkin, *Sapientia Astrologica*, 41–92.

65 Bacon, *Communia mathematica*, 40. Omnia que considerantur in Astrologia et Astronomia certificantur per vias geometricas et arismeticas, ut vult Ptholomeus in principio *Almagesti*, et per omnia que sunt in musicalibus manifestantur per vias numerorum et figurarum, ut planum est expertis Musice. Et ideo Geometria et Arismetica precedunt alias.

66 Bacon, *Communia mathematica*, 49.

67 Bacon, *Communia mathematica*, 50. Et non solum intendit certificari de celestibus, sed de omnibus que oriuntur supra orizonta, cuiusmodi sunt impressiones praetacte. Astronomo enim mathematico incumbit talia certificare quatenus tandem deveniat ad usum et experienciam cognoscendi futura, presencia et preterita, ut mala cognita impediatur et bona promoveantur, tam reipublice quam persone.

Bibliography

Primary Sources

Aristotle. *Meteorologies.* Edited by Gudrum Vuillemin-Diem. Turnhout: Brepols, 2008.

Petrus Peregrinus de Maricout. "Epistolae de Magnete/ Nova Compositio Astrolabii Particularis." In *Petrus Peregrinus de Maricourt, Opera*, edited by Loris Sturlese and Ron B. Thomson. Pisa: Scuola normale superiore, 1995.

Pseudo-Ptolemy. *Centiloquium.* Edited by Richard Lemay. Unpublished.

Ptolemy. *Almagest.* Translated by G. J. Toomer, in *Ptolemy's Almagest.* Princeton, NJ: Princeton University Press, 1986.

Ptolemy. *Tetrabiblos.* Edited and translated by Frank E. Robbins. Cambridge, MA and London: Harvard University Press, 1940.

Roger Bacon. *Communia Mathematica.* Edited by Robert Steele. *Opera hactenus inedita Rogeri Baconi*, fasc 16, Oxford: Clarendon, 1905–1940.

Roger Bacon. *Compendium studii philosophiae.* Edited by Thomas S. Maloney. Oxford: Oxford University Press, 2018.

Roger Bacon. "De speculis comburentibus." In *Roger Bacon's Philosophy of Nature*, edited by David C. Lindberg. Oxford: Oxford University Press, 1983.

Roger Bacon. "Epistola de secretis operibus artis et naturae et de nullitate magiae." In *Fr. Rogeri Bacon Opera Quaedam Hactenus Inedita, vol. I: Containing the Opus Tertium, Opus Minus, and Compendium Philosophiae*, edited by John S. Brewer. London: Longman, Green, Longman, and Roberts, 1859.

Roger Bacon. *Moral Philosophy.* Translated with Introduction and Notes by Thomas S. Maloney and Jeremiah Hackett. St. Bonaventure, NY: Franciscan Institute Press, 2022.

Roger Bacon. *Opus majus.* Edited by John Henry Bridges. *3 vols.* Oxford and Edinburgh: Clarendon, 1897–1900.

Roger Bacon. *Opus minus.* Edited by Thomas S. Maloney, forthcoming.

Roger Bacon. *Opus tertium* Edited and Translated by Nikolaus Egel. Hamburg: Felix Meiner Verlag, 2020.

Roger Bacon. *Opus tertium.* Edited by John S. Brewer. *Fr. Rogeri Bacon Opera Quaedam Hactenus Inedita, vol. I: Containing the Opus Tertium, Opus Minus, and Compendium Philosophiae.* London: Longman, Green, Longman, and Roberts, 1859.

Roger Bacon. *Opus tertium.* Edited by Andrew G. Little, *A Part of the Opus Tertium.* Aberdeen: The University Press, 1912.

Roger Bacon. *Secretum secretorum.* Edited by Robert Steele. *Opera hactenus inedita Rogeri Baconi*, fasc 5. Oxford: Clarendon, 1920.

Secondary Sources

Allen, Meagan S. *Roger Bacon and the Incorruptible Human (1220–1292): Alchemy, Pharmacology, and the Desire to Prolong Life.* London: Palgrave Macmillan, 2023.

Boudet, Jean-Patrice. "Astrology Between Rational Science and Divine Inspiration: The Pseudo Ptolemy's *Centiloquium.*" In *Dialogues among Books in Medieval Western Magic and Divination*, edited by Stefano Rapisarda and Erik Niblaeus, vol. 65. Florence: SISMEL: Edizioni Del Galuzzo, Micrologus Library, 2014.

Crisciani, Chiara. "'Experimentum,' ricetta, raccolta: Un percorso epistemologico." *Studi sull'Aristotelismo medievale (secolo VI–XVI)* 2/2 (December 2002): 223–242.

Duhem, Pierre. *Un Fragment Inédit de l'Opus Tertium de Roger Bacon.* Quarrachi, 1909.

Falcon, Andrea. *Aristotle and the Science of Nature: Unity without Unification.* Cambridge: Cambridge University Press, 2005.

Feke, Jacqueline. *Ptolemy's Philosophy: Mathematics as a Way of Life.* Princeton and Oxford: Princeton University Press, 2018.

Hackett, Jeremiah. "Ego Expertus Sum: Roger Bacon's Science and the Origins of Empiricism." In *Expertus sum. L'expérience par les sens dans la philosophie naturelle médiévale*, edited by Thomas Bénatouil and Isabelle Draelants, 145–173. Florence: SISMEL-Edizioni del Galluzzo, 2011.

Hackett, Jeremiah. "Roger Bacon on Experimental Science." In *Roger Bacon and the Sciences –Commemorative Essays,* edited by Jeremiah Hackett, 277–315. Leiden: Brill, 1997.

Hackett, Jeremiah. "Roger Bacon on Military Technology: A Brief Introduction." In *Geistige und körperliche Arbeit im Mittelalter,* edited by Günther Mensching and Alia Mensching-Estakhr, 123–135. Würtzberg: Könighausen und Neumann, 2010.

Hackett, Jeremiah. "Roger Bacon's Various Texts on Perspectiva in His Later Works: The Significance of Two New Versions of his Perspectiva." In *Arabic and Latin Theory of Perspective*, edited by Cecilia Panti, Agostino Parvicini Bagliani, and Nicholas Temple, 81–104. Florence: SISMEL-Edizioni Del Galuzzo, Micrologus 24, 2021.

Hackett, Jeremiah. "A Working Edition of Roger Bacon: Tractatus in qui fit sermo de experiential in communi." In *The Meaning of Experimental Science (scientia experimentalis) in the Philosophy of Roger Bacon*, edited by Jeremiah M. G. Hackett. University of Toronto PHD, 1983.

Kedar, Yael and Giora Hon. "Roger Bacon (c. 1220–1292) and his System of Laws of Nature: Classification, Hierarchy and Significance." *Perspectives on Science* 25/6 (2017): 719–745.

Lindberg, David C. "Roger Bacon on Light, Vision, and the Universal Emanation of Force." In *Roger Bacon and the Sciences – Commemorative Essays,* edited by Jeremiah Hackett, 243–275. Leiden: Brill, 1997.

Nagel, Silvia. "Pietro Peregrino, il sapiens-simplex excellente di Ruggero Bacone." In *I Francescani e le scienze*, 21–47. Spoleto: CISAM, 2012.

Panti, Cecilia. "The Theological Use of Science in Robert Grosseteste and Adam Marsh According to Roger Bacon. The Case Study of the Rainbow." In *Robert Grosseteste and the Pursuit of Religious and Scientific Learning in the Middle*

Ages, edited by Jack Cunningham and Mark Hocknull, 143–163. Berlin: Springer, 2016.

Rutkin, H. Darrel. "An Idealized Astrological Courtier at a 13th Century Papal and Royal Court: The Case of Roger Bacon." In *De Frédéric II á Rodolphe II: Astrologie, divination et magie dans les cours XIIIe-XVIIe siècle*, edited by Jean-Patrice Boudet, Martine Ostorero, and Agostino Paravicini Bagliani, 53–69. Florence: SISMEL Edizioni Del Galuzzo: Micrologus Library 85, 2017.

Rutkin, H. Darrel. *Sapientia Astrologica: Astrology, Magic and Natural Knowledge, ca, 1250 1800, I Medieval Structures (1250–1500)*. Springer, 2019.

Weill-Parot, Nicolas. *Les "images astrologiques" au Moyen Âge et à la Renaisance: Spéculations intellectuelles et practiques magiques (XIIe–XVe siècles)*. Paris: Honoré Champion Éditeur, 2002.

2 The Division of the Sciences in *Opus maius* Part IV, 1, 2

Roger Bacon's Sources and Their Transformation

Alexander Fidora

2.1 Introduction

Scholarship on Roger Bacon has repeatedly emphasised the importance of mathematics within his thought and, more specifically, within his epistemology. While Bacon did not delve into mathematical theory as such, he was extremely interested in mathematics and in its usefulness for the other sciences. In fact, he claimed that mathematics was key to understanding all other disciplines.[1]

This epistemological interest in mathematics was not exclusive to Bacon. Rather, it indicates his participation in a wider debate of his time regarding the certainty of the speculative sciences, namely, natural philosophy, mathematics, and metaphysics. In the 1230s, Richard Rufus brought up the question in his *Memoriale quaestionum in Metaphysica Aristotelis* of whether mathematics or metaphysics should be considered the most certain science.[2] That the issue was controversial among Bacon's contemporaries is confirmed by a remark in Robert Kilwardby's *De ortu scientiarum*, from approximately 1250. In Chapter 63 of the work, Kilwardby outlines the following scenario:

> In this order, wherein physics undoubtedly succeeds metaphysics and mathematics, one person might say that mathematics precedes metaphysics because of the most certain demonstrations that obtain in the former. A different person, however, might claim the opposite, since metaphysics is the supreme part of philosophy, and therefore its principles must be self-evident and have the power to explain the common principles of the particular sciences. And perhaps each of these statements is true, namely, that mathematics is more certain than and prior to metaphysics as regards its demonstrations, while metaphysics is more certain than and prior to mathematics as regards its explanations and elucidations of the other sciences.[3]

In this chapter, I wish to analyse Bacon's approach to this question as laid out in *Opus maius* IV, 1. In this Distinction, he devotes two chapters to the epistemological status of mathematics: Chapter 2: "In quo probatur per auctoritatem, quod omnis scientia requirit mathematicam," and Chapter 3: "In

DOI: 10.4324/9781003405702-3

quo probatur per rationem, quod omnis scientia requirit mathematicam."[4] While the latter could seem more promising for a systematic analysis of Bacon's own position, I will focus on the former, as it offers crucial insights into Bacon's reception and transformation of the traditional division of the sciences, which reception and transformation will be at the centre of this chapter.

2.2 Essential vs Accidental Modes of Philosophy

Chapter 2 of *Opus maius* IV, 1 builds on a macro-division of philosophy into what is called its essential and accidental parts, in an attempt to show that every discipline contained in both of these parts depends on mathematics. With regard to the essential parts, Bacon states the following: "For since there are three essential parts of philosophy, as Aristotle says in the sixth book of his *Metaphysics*, mathematics, natural and divine, the mathematical is of no small importance in grasping the knowledge of the two other parts."[5]

For the essential parts of philosophy, Bacon here refers to Aristotle's classification of theoretical philosophy in *Metaphysics* VI, which is indeed the *locus classicus* of the division of theoretical philosophy into physics, mathematics, and metaphysics. However, Aristotle's original text does not speak of essential modes of philosophy. It was Michael Scot in his Arabic-Latin translation of *Metaphysics* VI, 1 who applied the expression "modi essentiales" to the three parts of Aristotle's theoretical philosophy.[6] Yet no reference whatsoever is found in any of Aristotle's works, whether in Greek or in Latin translation, to a corresponding accidental mode of philosophy.

The idea of an accidental part of philosophy, which Bacon uses when he goes on to claim that "the accidental parts of philosophy are grammar and logic,"[7] is not, in fact, integral to the Aristotelian tradition. Rather, this doctrine is characteristic of the introductions to philosophy from the first half of the thirteenth century. These were propaedeutic treatises, written for the purpose of teaching, that offered definitions of philosophy along with a division of its parts and an account of the relations between these parts. The introductions to philosophy were for a long time neglected by traditional scholarship. However, the editions by Claude Lafleur and others of a series of such texts from the first half of the thirteenth century have confirmed their significance,[8] insofar as the texts allow us to assess the self-understanding of the philosophical masters and of their teaching during the first decades of the University of Paris.

Elaborating upon Aristotle's *Metaphysics*, the authors of these introductions complemented the division of essential philosophy introduced by Michael Scot by supplying the corresponding category of its accidental parts. The anonymous Parisian introduction *Sicut recitat Avicenna*, from the middle of the thirteenth century, explains the case as follows: "As Aristotle says that there are three essential modes of knowing, that is, the natural, the mathematical and the divine one, he seems to imply that the other three are accidental, namely, grammar, logic and rhetoric."[9]

Bacon's bipartite division of philosophy into essential and accidental parts, which provides the general structure for *Opus maius* IV, 1, 2, clearly reflects this well-established view at the Paris Faculty of Arts.[10]

2.3 Mathematics and the Essential Modes of Philosophy

Starting his discussion of the relation between mathematics and the other essential modes of philosophy, that is, natural philosophy and divine science, Bacon first invokes the authority of Ptolemy, claiming that mathematics is crucial for both natural philosophy and divine science, "as Ptolemy teaches in the first chapter of the *Almagest.*"[11] The passage Bacon refers to – without, however, quoting it – comes from *Almagest* I, 1:

> The first two divisions of theoretical philosophy should rather be called guesswork (*eikasia*) than knowledge, theology because of its completely invisible and ungraspable nature, physics because of the unstable and unclear nature of matter; hence there is no hope that philosophers will ever be agreed about them; and that only mathematics can provide sure and unshakeable knowledge to its devotees, provided one approaches it rigorously.[12]

According to Bacon, this passage establishes the fundamental role of mathematics vis-à-vis physics and metaphysics. The same view was entertained by other thirteenth-century authors, who argued on very similar lines in favour of mathematics: for instance, Thomas Aquinas in his commentary on Boethius's *De trinitate*, q. VI, a. 1. In this text, written between 1255 and 1259, during which period Aquinas became regent master at Paris, one finds the following account in reference to mathematics:

> The method of mathematics is also more certain than the method of divine science, because the objects of divine science are further removed from sensible things, from which our knowledge takes its origin [....] But mathematical entities do fall under the senses and they are objects of our imagination, for example, figures, lines, numbers, and the like [....] It is clear, then, that mathematical inquiry is easier and more certain than physical and theological [....] This is what Ptolemy asserts in the beginning of the *Almagest*: "Let us call the other two kinds of theoretical knowledge opinion rather than science: theology because of its obscurity and incomprehensibility, physics because of the instability and obscurity of matter. The mathematical type of investigation alone will give the inquirer firm and unshaken certainty through demonstrations carried out by unquestionable methods."[13]

Bacon and Aquinas seem to agree on the privileged role of mathematics, which they both illustrate with a text from Ptolemy.

From here, Bacon moves rather unexpectedly to practical philosophy. At this juncture, Bacon's argument becomes very interesting; for what reason – one may ask – does Bacon discuss practical philosophy among the essential modes of philosophy, which actually correspond to Aristotle's theoretical disciplines? The reason for him to do so is the twofold nature of divine science:

> [D]ivine science is twofold, as is clear from the first book of the *Metaphysics*, namely, the first philosophy, which shows that God exists, whose exalted properties it investigates, and civil science, which determines divine worship, and explains many matters concerning God as far as a man can receive them.[14]

Basing himself on Aristotle's *Metaphysics* I, 2, 983a 5, Bacon claims that divine science embraces both first philosophy or metaphysics *sensu stricto* and civil science, which concerns the divine cult. If mathematics is fundamental for divine science, it must be so for both its theoretical and its practical parts. It is in this context that Bacon quotes at length – and not only by allusion as in the case of Ptolemy – a passage from Boethius' *De institutione arithmetica* II, Chapter 45, which allows him to establish the connection between mathematics and politics:

> Hence Boethius says at the end of his *Arithmetic* that "the mathematical means are found in civil polity. For he says that an arithmetic mean is comparable to a state that is ruled by a few, for this reason, that in its lesser terms is the greater proportion; but he states that there is a harmonic mean in an aristocratic state for the reason that in the greater terms the greater proportionality is found. The geometrical mean is comparable to a democratic state equalised in some manner; for whether in their lesser or greater terms they are composed of an equal proportion of all. For there is among all a certain parity preserving a law of equality in their relations."[15]

I will not expand on this text, which offers an interesting reflection on the various forms of statehood in terms of proportion or participation, as one would say today. Rather, I would like to draw attention to the Aristotelian passage which allows Bacon to introduce the practical disciplines, and more specifically politics, into the essential modes of philosophy. In fact, *Metaphysics* I, 2 983a 5–7 is itself obscure, allowing for various interpretations. It suggests that divine science is called so in two ways, namely with regard to that science or knowledge which belongs to God alone and to that which has God as its subject of study.[16] It is worth noting that Thomas Aquinas also referred to this passage in his commentary on Boethius's *De trinitate*, proposing, however, a different reading of the text, namely as grounding his distinction between revealed theology on the one hand and metaphysics on the other.[17] In contrast, Bacon's interpretation of the passage is close to some of his Arabic sources, and in particular to Avicenna's *Metaphysics*. The final chapters of this text, which play an important role in Bacon's *Philosophia moralis*, establish a systematic link between divine science on the one hand and ethics and politics on the other,[18] as other Arabic authors, such as al-Fārābī, had done before in their divisions of the sciences.[19]

As a result of his reading of Aristotle's *Metaphysics*, Bacon substantially transforms the category of essential philosophy, which the early masters of philosophy identified with Aristotle's speculative sciences. While it originally served to classify the different branches of theoretical philosophy, it now includes practical philosophy as well. That this transformation is not accidental is confirmed by another text by Bacon: to wit, his *Summulae dialectices* from the 1250s. In this text, which is strongly dependent on Parisian teachings, Bacon briefly presents the division into essential and accidental modes of philosophy, which he describes as follows:

> According to some, grammar, logic, and rhetoric are parts of the trivium and accidental modes of philosophy, the latter of these being concerned with the signs of things, such as speech. Natural philosophy, mathematics and metaphysics and moral philosophy are essential modes and they are the principal parts of the quadrivium.[20]

This passage corroborates the idea that Bacon was very serious about including practical or moral philosophy among the essential modes. Although this inclusion would clearly run counter to the standard division of philosophy in the thirteenth century, one can identify at least one text that could have inspired Bacon: that is, the anonymous introduction to philosophy *Philosophica disciplina*, from circa 1245:

> In a different way the speculative philosophy of things is divided by Aristotle in Book VI of his *Metaphysics*, where he says that philosophy has three essential modes: natural philosophy, mathematics, and divine science, the other mode being accidental. Divine science is twofold, as Aristotle insinuates at the beginning of his *Metaphysics*: either it deals with God and the divine substances, like metaphysics, or it deals with that which disposes humans correctly and allows them to rejoin divine goodness and rejoice in it – "disposing" concerns the virtues and the happiness in this life, while "rejoining" refers to beatitude and happiness in the other life [....] The accidental mode of philosophy, in turn [...,] refers to the speech-related disciplines.[21]

Like Bacon, the anonymous author places practical philosophy among the essential modes of philosophy, drawing on the same text from Aristotle's

Metaphysics I, 2 that we have come across in Bacon.[22] It must be noted that this position is quite unique and was clearly opposed by other authors. Arnulf of Provence, for instance, explicitly held the opposite view in his *Divisio scientiarum* (ca. 1250):

> The proper mode of dividing philosophy according to Aristotle in book six of his *Metaphysics* is as follows: Among the modes of philosophy, one is essential, the other accidental. There are three essential modes, namely natural philosophy, mathematics and divine science. The accidental mode of philosophy concerns everything which has to do with human affairs, like speech and the virtues and vices.[23]

Arnulf unmistakably counts practical philosophy among the accidental modes of philosophy; in all other philosophical introductions that I have had access to, the authors leave practical philosophy to one side, considering it a branch *sui generis*.

Against the background of the thirteenth-century introductions to philosophy, there can be no doubt that Bacon's division of the sciences is very much indebted to the Parisian arts masters. Yet it is no less true that the English Franciscan did not simply follow the mainstream but pushed for a reorganisation of the essential modes of philosophy, so as to integrate practical philosophy along with metaphysics. Whether inspired by his Arabic sources or, more directly, by the *Philosophica disciplina* and its peculiar notion of divine science, this reinterpretation – which is not devoid of systematic tensions – must be considered characteristic of Bacon's overall thought.[24]

2.4 Mathematics and the Accidental Modes of Philosophy

In his analysis of the relation between mathematics and the accidental parts of philosophy, which occupies the second part of *Opus maius* IV, 1, 2, Bacon refers his readers no fewer than four times to al-Fārābī and his treatise *De scientiis*. This treatise, which exerted a strong influence during the thirteenth century, was translated by Dominicus Gundissalinus and also by Gerard of Cremona in Toledo. According to Bacon, "al-Fārābī makes it clear in his book *On the Sciences* that grammar and logic cannot be known without mathematics."[25] Although Bacon does not quote a specific passage from the work, the following lines from its chapter on logic may be at the origin of his argument:

> And this art [i.e. logic] is related to the art of grammar. And this is so because the relation between logic and reason viz. the objects of reason is like the relation between grammar and language viz. the words [....] This art is also related to the science of *alhorod*. In fact, the relation between the science of logic and the intelligibles is like the relation between *alhorod* and the rhythms of a verse. Similarly to all the rules

> that the science of *alhorod* offers us regarding the quantity of verses, the science of logic offers us rules for the objects of reason. Further, the rules of logic, which are instruments by means of which one tests rational contents with regard to which reason has probably already committed a mistake [...] resemble the weights and measures, which are instruments by means of which one tests several bodies with regard to which the senses have probably already committed a mistake [...] they are like the rules by means of which one tests the lines regarding which the senses have probably already committed a mistake.[26]

This passage, which is only included in Gerard's rendering of the *De scientiis*, draws several analogies between logic, grammar, the science of *alhorod* (i.e., prosody), and the science of weights, measures, and geometry. Bacon offers his own particular reading of these analogies, as he seems to understand them as systematic interdependences rather than analogies. Thus, focusing on the analogy between logic and prosody, Bacon claims that the musico-mathematical categories of metre and rhythm are fundamental for both grammar and logic, which are described as subordinate disciplines: "Therefore grammar depends causatively on music. In the same way logic." [27]

While this is a somewhat forced interpretation of al-Fārābī, Alain Galonnier has pointed to a paragraph in the poetic chapter of *De scientiis* which does establish a systematic link between poetical language, with its metres and rhythms, and logic.[28] Here, al-Fārābī holds that in order to move one's soul to action, poetical language needs not only to be beautiful, but also to incorporate elements from logic:

> For this reason [that is, to move one's soul to action] poetical speech, in contrast to all others, is endowed with beauty, grace and subtlety, and with elegance and clarity by means of all that is said in the science of logic.[29]

But again, Bacon's argument has a different emphasis; in his understanding, it is not poetics that is in need of logic so as to be able to achieve its perfection, but rather logic finds its perfection in poetical and rhetorical language, since the "ultimate goal of logic is the composition of arguments which move the practical intellect to faith and to a love of virtue and future felicity."[30] To this end, Bacon says, logic has to strive for the beauty of language, which al-Fārābī described in his chapter on poetics, along with metres and rhythms.

On the whole, and in spite of his emphatic references to al-Fārābī, Bacon's use of the *De scientiis* is rather idiosyncratic as it offers a very personal interpretation of *De scientiis* in an attempt to argue for the dependence of grammar and logic on mathematics. Key to understanding this interpretation is, once again, practical philosophy. According to Bacon, logic has to move the practical, and not the theoretical, intellect, and for this reason, it needs poetics and rhetoric, which in turn are based on musico-mathematical notions. This reasoning entails a significant extension of the ends of Aristotle's poetics and rhetoric to the realm of logic. The reappraisal of poetics and rhetoric, characteristic of Bacon's thought, also explains why in *Opus maius* IV, 1, 2 rhetoric is not mentioned as it appears in other contemporary divisions: that is, as a distinct science contained among the accidental mode of philosophy along with grammar and logic. As Jeremiah Hackett has argued, such a "division reduced the truth bearing capacity of rhetoric and poetics," opposing them to grammar and logic.[31] Instead, Bacon advocated an integrated concept of grammar, rhetoric, logic, and poetics which was orientated towards practical or moral philosophy.

That the omission of rhetoric in Bacon's account of the accidental modes of philosophy is not incidental is confirmed by the following passage in the *Opus tertium*: "It is clear that logic and grammar are accidental sciences, and not the principal ones."[32] While in his early *Summulae dialectices* Bacon did count rhetoric among the accidental modes of philosophy,[33] the *Opus maius* and the *Opus tertium* quite deliberately avoid mentioning this standard division. In Bacon's account, rhetoric and poetics, rather than being concerned with "the non-logical (non-truthful) aspect of speech,"[34] provide the link between the accidental modes of philosophy and the highest mode of essential philosophy: that is, practical or moral philosophy. Consequently, in Bacon a proper discussion of rhetoric and poetics is found not among the modes of accidental philosophy, but in the seventh section of the *Opus maius*, that is, the *Moralis philosophia*.[35]

2.5 Conclusion

This close analysis of *Opus maius* IV, 1, 2 and its sources has shown how Roger Bacon transformed Latin and Arabic teachings in a very creative manner. Discussing the epistemological status of mathematics, he developed his own division of the sciences within the framework of some of the most influential epistemological accounts of his day: the introductions to philosophy by the Parisian arts masters and al-Fārābī's treatise *De scientiis*. While Bacon took inspiration from these texts, his emphasis on the practical or moral dimensions of the various parts of philosophy must be considered his personal imprint. As a result of this practical turn in his epistemology, Bacon adds one discipline to the three essential modes of philosophy while omitting another one from the three accidental modes.

Most striking is, without doubt, Bacon's inclusion of practical philosophy among the essential modes of philosophy, which contrasts with the mainstream of contemporary introductions to philosophy, although, as we have noted, one such introduction, namely the anonymous *Philosophica disciplina*, puts forward a similar approach. But not less telling is Bacon's omission of rhetoric from the list of the accidental modes of philosophy. Also, in this context, one can observe the crucial role of practical philosophy in providing the ultimate horizon for grammar and logic, whose goal is to move the practical intellect. For this purpose, they need to be perfected by rhetoric and poetics, which Bacon considers to be parts of practical philosophy.

It is certainly true, as has been noted many times, that Bacon's curricular programme launches a harsh critique against contemporary academic learning;[36] yet it is not disconnected from the teachings at the University of Paris. Rather, the foregoing analysis has shown how some of the most salient features of Roger Bacon's epistemology, which shaped his overall thought – such as the rapprochement of moral philosophy and metaphysics or the reappraisal of rhetoric and poetics – resulted from a productive dialogue with the philosophical discourse of his day.

Notes

1 See, for instance, Frankowska-Terlecka, "'Scientia' as Conceived by Roger Bacon," 209–231; Hackett, "Roger Bacon on the Classification of the Sciences," 49–65; Hackett, "Roger Bacon on Rhetoric and Poetics," as well as the chapter by Jeremiah Hackett in this volume.

2 Rufus, *Memoriale quaestionum in Metaphysica Aristotelis*. An metaphysica sit scientia certissima. De tertio: videtur enim esse certissima, cum sit de causis certissimis; est enim de causa omnium prima quae certissima est. Contrarium videtur – certior est scientia quae procedit ex principiis quam quae est ad principia; haec autem est ad principia – nisi quod hic solvit prior distinctio.

3 Kilwardby, *De ortu scientiarum*, 218. In hoc autem ordine cum absque dubio physica sit posterior metaphysica et mathematica, aliquis tamen diceret mathematicam esse priorem metaphysica propter certissimas demonstrationes quae sunt in mathematica. Aliquis forte diceret e converso eo quod metaphysica est suprema philosophiae pars, et ideo eius principia oportet esse per se notissima et aliorum communium principiorum specialium scientiarum explanativa. Et forte utrumque verum est, scilicet quod mathematica est certior et prior in demonstrando, sed metaphysica certior et prior in explanando et alias declarando.

4 Bacon, *Opus Maius*, vol. I, 98–103 and 103–108, respectively.

5 English translation by Robert Belle Burke, vol. 1, 117. Original text Bacon, *Opus Maius*, vol. 1, 98–99. Cum enim sint modi tres philosophiae essentiales, ut dicit Aristoteles in sexto *Metaphysicae*, mathematicus, naturalis et divinus, non parum valet mathematicus ad reliquorum duorum modorum scientiae comprehensionem.

6 Michael Scot's translation of Aristotle, *Metaphysics* VI, 1, 1026a 18–19 reads: Modi igitur philosophiae essentiales sunt tres, mathematicus, naturalis et divinus. *Aristotelis Metaphysicorum libri XIII, fol. 145 L-M.*

7 Bacon, *Opus Maius*, vol. I, 99 (118). Modi autem philosophiae accidentales sunt grammatica et logica.

8 See in particular Lafleur, *Quatre Introductions à la philosophie au* XIIIe *siècle*; and Lafleur, *L'enseignement de la philosophie au* XIIIe *siècle*.

9 Edited in Lafleur and Carrier, "La *Philosophia* d'Hervé le Breton," 359–442, here 39. Cum enim dicat Aristoteles tres esse essentiales modos scientiae, naturalem scilicet, mathematicum et divinum, in hoc innuit tres esse alios accidentales, scilicet grammaticum, logicum et rhetoricum.

10 The same division occurs, e.g., in the anonymous introduction *Ut ait Tullius*, edited by Dahan, "Une introduction à l'étude de la philosophie," 3–58; Oliverus Brito, *Philosophia*, in Lafleur and Carrier, L'introduction à la philosophie de maître Olivier, 467–487, here p. 482; and also Robertus Anglicus, *Isagoge*, in Piché, Le problème des universaux, p. 265.

11 Bacon, *Opus Maius*, vol. I, 99 (117). Ut docet Ptolemaeus in capitulo primo *Almagesti*.

12 Ptolemy, *Almagest*, 36. For a discussion of the introduction of the *Almagest* and its various Latin renderings see Burnett, "Ptolemaeus in Almagesto dixit," 115–140.

13 Thomas Aquinas, *The Division and Methods of the Sciences*, 68–69. Est etiam processus mathematicae certior quam processus scientiae divinae; quia ea, de quibus est scientia divina, sunt magis a sensibilibus remota, a quibus nostra cognitio initium sumit [....] Mathematica autem ipsa in sensu cadunt et imaginationi subiacent, ut figura, linea et numerus, et huiusmodi [....] Et sic patet quod mathematica consideratio est facilior et certior quam naturalis et theologica, [...] Et hoc est quod Ptolemaeus dicit in principio *Almagesti*: "Alia duo genera theorici potius quis opinionem quam conceptionem scientialem dicat, theologicum quidem propter inapparens ipsius et incomprehensibile, physicum vero propter materiae instabile et immanifestum; solum autem mathematicum inquisitionis firmam stabilemque fidem intendentibus dabit, velut utique demonstratione per indubitabiles vias facta" (Ed. Leon. L, p. 161).

14 Bacon, *Opus Maius*, vol. I, 99 (117). [D]ivinus sit dupliciter, ut patet ex primo *Metaphysicae*, scilicet philosophia prima, quae Deum esse ostendit, cuius proprietates excelsas investigat, et civilis scientia, quae cultum divinum statuit, multaque de eo secundum possibilitatem hominis exponit, ad utramque istarum multum valere mathematicam.

15 Bacon, *Opus Maius*, vol. I, 99 (118, Burke's translation slightly modified). Unde Boethius in fine *Arithmeticae* mathematicas medietates asserit in rebus civilibus inveniri. Dicit enim quod "arithmetica medietas reipublicae comparatur quae a paucis regitur, idcirco quod in minoribus eius terminis maior proportio fit, musicam vero medietatem optimatum dicit esse rempublicam, eo quod in maioribus terminis maior proportionalitas invenitur. Geometrica medietas popularis quodammodo exaequatae civitatis est: namque vel in minoribus vel in maioribus aequali omnium proportionalitate componuntur. Est enim inter omnes paritas quaedam medietatis aequum ius in proportionibus conservantis." Compare Boethius's *De arithmetica*: Atque ideo arithmetica quidem rei publicae comparatur, quae paucis regitur, idcirco quod in minoribus eius terminis maior proportio sit. Musicam vero medietatem optimatium dicunt esse rem publicam ideo, quod in maioribus terminis maior proportionalitas invenitur. Geometrica medietas popularis quodammodo et exaequatae civitatis est. Namque vel in maioribus vel in minoribus aequali omnium proportionalitate componitur, et est inter omnes paritas quaedam medietatis aequum ius in proportionibus conservantis. In Boethius, *De arithmetica*, 191.

16 See Aristotle, *Metaphysics. Books I-IX*, 15. A science is divine if it is peculiarly the possession of God, or if it is concerned with divine matters.

17 See Thomas Aquinas, *Super Boethium De trinitate* q. II, a. 2, c. Sed divinorum notitia dupliciter potest aestimari: uno modo ex parte nostra, et sic nobis cognoscibilia non sunt nisi per res creatas, quarum cognitionem a sensu accipimus; alio modo ex natura ipsorum, et sic ipsa sunt ex se ipsis maxime cognoscibilia, et quamvis secundum modum suum non cognoscantur a nobis, tamen a Deo cognoscuntur et a beatis secundum modum suum. Et secundum hoc de divinis duplex scientia habetur: una secundum modum nostrum, qui sensibilium principia accipit ad notificandum divina, et sic de divinis philosophiae scientiam tradiderunt, philosophiam primam scientiam divinam dicentes; alia secundum modum ipsorum divinorum, ut ipsa divina secundum se ipsa capiantur [....] The text is followed by a reference to *Metaphysics* I, 2, 983a 5–7: Propter quod [sapientia] etiam dea scientiarum dicitur in principio *Metaphysicae*, et multa magis haec quae non solum de altissimis sed ex altissimis est.

18 See Avicenna, *Liber de philosophia prima sive scientia divina*, vol. 2, 542–553.

19 See note 35.

20 de Libera, "Les *Summulae dialectices,*" 172. Vel secundum alios grammatica, logica, rhetorica sunt partes trivii et modi philosophiae accidentales, quae sunt de signis rerum, sicut de sermone. Naturalis, mathematica et mechanica [*leg.* metaphysica] et moralis sunt essentiales, quia sunt de rebus, et sunt partes principales quadrivii.

21 Lafleur, *Quatre Introductions à la philosophie au* XIIIe *siècle*, 261–262. Aliter dividitur philosophia rerum speculativa ab Aristotele sexto *Metaphysicae*, ubi dicit quod modi essentiales philosophiae sunt tres: naturalis, mathematicus et divinus; alius accidentalis. Divinus duplex est, sicut innuit Aristoteles in principio *Metaphysicae*: alius enim considerat de Deo et de substantiis divinis, sicut metaphysica; alius est de eis quae ordinant hominem et coniungunt bonitati divinae fruendae – ordninant autem sicut virtutes et felicitas temporalis quae habetur in hac vita; coniungunt sicut beatitudo sive felicitas patriae [....] Accidentalis autem modus philosophiae [...] est sermocinalis.

22 Cecilia Panti and Roberto Lambertini have drawn attention to further similarities between the *Philosophica disciplina* and other introductions to philosophy on the one hand and Roger Bacon's works on the other. See Panti, "Sound, Light and Cosmic Music," 367–383, at 380–381; Lambertini, "*Tota familia Aristotelis,*" 130–131.

23 Lafleur, *Quatre Introductions à la philosophie au* XIIIe *siècle*, 315. Modus proprius dividendi philosophiam secundum Aristotelem sexto *Methaphysicae* est iste: modorum philosophiae, unus est essentialis, alius accidentalis. Modi philosophiae essentiales sunt tres: naturalis, mathematicus et divinus. Accidentalis vero modus philosophiae est omnis qui est de rebus humanis, ut sunt sermones, virtutes et vitia. Horum enim sunt modi philosophiae accidentales.

24 See Wieland, "Ethik und Metaphysik," 147–173.

25 Bacon, *Opus Maius*, vol. I, 99 (118, Burke's translation slightly modified). Et quod sine mathematica non possunt sciri scientiae istae patet per Alpharabium in libro *De scientiis.*

26 Galonnier, *Le "De scientiis Alfarabii" de Gérard de Crémone,* 180–182. Et haec quidem ars proportionatur arti grammaticae. Et illud est quia proportio artis dialecticae ad rationem et rationata est sicut proportio artis grammaticae ad linguam et dictiones [....] Et proportionatur iterum scientiae *alhorod*: proportio enim scientiae dialecticae ad rationata est sicut proportio *alhorod* ad ponderum versuum. Et totum quod nobis dat scientia *alhorod* ex canonibus in ponderibus versuum eius compar dat nobis scientia dialecticae in rationatis. Et iterum canones dialecticae, qui sunt instrumenta quibus experitur in rationatis de quibus non est securitas quin ratio iam erraverit in eis [...,] sunt similes ponderibus et mensuris, quae sunt instrumenta cum quibus experitur in pluribus corporum de quibus non est securitas quin sensus iam erraverit [...] et sicut regulae quibus experitur in lineis de quibus non existit securitas quin sensus iam erraverit.

27 Bacon, *Opus Maius*, vol. I, 100 (119). Ergo grammatica dependet causaliter ex musica. Eodem modo loigica.

28 Galonnier, *Le "De scientiis Alfarabii" de Gérard de Crémone*, 131–132.

29 Galonnier, *Le "De scientiis Alfarabii" de Gérard de Crémone*, 212. Et propter illud facti sunt sermones poetici absque aliis pulchri et decentes et sublimes, et ponuntur eis decor et delcaratio cum rebus quae dictae sunt in scientia logicae.

30 Bacon, *Opus Maius*, vol. I, 100 (119, Burke's translation slightly modified). Finis logicae est compositio argumentorum quae movent intellectum practicum ad fidem et amorem virtutis et felicitatis futurae.

31 Hackett, "Roger Bacon on Rhetoric and Poetics," 149.

32 Bacon, *Opus tertium*, 105. Et ideo patet quod logica et grammatica sunt accidentales scientiae, et non principales [....] On this passage, see also Perler, "Logik – eine 'wertlose Wissenschaft,'" 379.
33 See note 20.
34 Hackett, "Roger Bacon on Rhetoric and Poetics," 149.
35 In this context, al-Fārābī's *De scientiis* remains an important source of inspiration, particularly Chapter 5, in which al-Fārābī (or rather his Latin translator Gerard) presents an "ars eloquentiae" (that is Gerard's rendering of *kalam*) along with ethics and politics. See Rosier-Catach, "Roger Bacon," 87–110.
36 See, e.g., Egel, "Roger Bacon's New Founding of the Sciences," 125–140.

Bibliography

Primary Sources

Aristotle. *Aristotelis Metaphysicorum libri XIII. Cum Averrois Cordubensis in eosdem commentariis.* Venice: Apud Iunctas, 1562.

Aristotle. *Metaphysics, Books I–IX.* Translated by Hugh Tredennick. Cambridge, MA and London: Harvard University Press/Heinemann, 1933.

Avicenna. *Liber de philosophia prima sive scientia divina.* Edited by Simone van Riet, *2 vols.* Louvain and Leiden: Peeters/Brill, 1977–1980.

Boethius. *De arithmetica.* Edited by Henri Oosthout and Johannes Schilling. Turnhout: Brepols, 1999.

Ptolemy. *Almagest.* Translated by G. J. Toomer, in *Ptolemy's Almagest.* Princeton: Princeton University Press, 1986.

Richard Rufus. *Memoriale quaestionum in Metaphysica Aristotelis.* Edited by Rega Wood and Neil Lewis, 2013. <https://rrp.stanford.edu/MMet.shtml>.

Robert Kilwardby. *De ortu scientiarum.* Edited by Albert G. Judy. Oxford: The British Academy/Pontifical Institute of Mediaeval Studies, 1976.

Roger Bacon. *Opus majus.* Edited by John Henry Bridges. *3 vols.* Oxford and Edinburgh: Clarendon, 1897–1900.

Roger Bacon. *The Opus Maius.* Translated by Robert Belle Burke. *2 vols.* Philadelphia: University of Pennsylvania Press, 1928.

Roger Bacon. *Opus tertium.* Edited by John S. Brewer. *Fr. Rogeri Bacon Opera Quaedam Hactenus Inedita, vol. I: Containing the Opus Tertium, Opus Minus, and Compendium Philosophiae.* London: Longman, Green, Longman, and Roberts, 1859.

Thomas Aquinas. *The Division and Methods of the Sciences. Questions V and VI of his Commentary on the De Trinitate of Boethius.* Translated by Armand Maurer. Toronto: Pontifical Institute of Mediaeval Studies, 1986.

Secondary Sources

Burnett, Charles. "'Ptolemaeus in Almagesto dixit': The Transformation of Ptolemy's Almagest in its Transmission via Arabic into Latin." In *Transformationen antiker Wissenschaften*, edited by Georg Toepfer and Hartmut Böhme, 115–140. Berlin and New York: de Gruyter, 2010.

Dahan, Gilbert. "Une introduction à l'étude de la philosophie: Ut ait Tullius." In *L'enseignement de la philosophie au XIIIe siècle. Autour du 'Guide de l'étudiant' du ms. Ripoll 109*, edited by Claude Laufleur and Joanne Carrier, 3–58. Turnhout: Brepols, 1997.

De Libera, Alain, "Les *Summulae dialectices* de Roger Bacon I-II. *De termino*, De enuntiatione." *Archives d'histoire doctrinale et littéraire du Moyen Âge* 53 (1989): 139–289.

Egel, Nikolaus. "Roger Bacon's New Founding of the Sciences in the Thirteenth Century." In *The Philosophy and Science of Roger Bacon. Studies in Honour of Jeremiah Hackett*, edited by Nicola Polloni and Yael Kedar, 125–140. London and New York: Routledge, 2021.

Frankowska-Terlecka, Małgorzata. "'Scientia' as Conceived by Roger Bacon." *Organon* 6 (1969): 209–231.

Galonnier, Alain. *Le 'De scientiis Alfarabii' de Gérard de Crémone. Contribution aux problèmes de l'acculturation au XIIe siècle*. Turnhout: Brepols, 2016.

Hackett, Jeremiah. "Roger Bacon." In *The Stanford Encyclopedia of Philosophy*, edited by Edward N. Zalta. (Summer 2020 Edition). <https://plato.stanford.edu/archives/sum2020/entries/roger-bacon/>

Hackett, Jeremiah. "Roger Bacon on the Classification of the Sciences." In *Roger Bacon and the Sciences – Commemorative Essays,* edited by Jeremiah Hackett, 49–66. Leiden: Brill, 1997.

Hackett, Jeremiah. "Roger Bacon on Rhetoric and Poetics." In *Roger Bacon and the Sciences – Commemorative Essays,* edited by Jeremiah Hackett, 133–149. Leiden: Brill, 1997.

Lafleur, Claude, *Quatre Introductions à la philosophie au XIIIe siècle. Textes critiques et étude historique*. Montréal and Paris: Institut d'Études Médiévales/Vrin, 1988.

Lafleur, Claude and Joanne Carrier. "L'introduction à la philosophie de maître Olivier le Breton." In *L'enseignement de la philosophie au XIIIe siècle. Autour du 'Guide de l'étudiant' du ms. Ripoll 109*, edited by Claude Lafleur and Joanne Carrier, 467–487. Turnhout: Brepols, 1997.

Lafleur, Claude and Joanne Carrier. "La *Philosophia* d'Hervé le Breton (alias Henri le Breton) et le recueil d'introductions à la philosophie du ms. Oxford, Corpus Christi College 283 (deuxième partie)." *Archives d'histoire doctrinale et littéraire du Moyen Âge* 62 (1995): 359–442.

Lambertini, Roberto. "*Tota familia Aristotelis.* On Some Sources of Bacon's Contribution to Medieval Political Discourse." *Revista española de filosofa medieval* 28/1 (2021): 125–147.

Panti, Cecilia. "Sound, Light and Cosmic Music: Grosseteste's Commentary on *Posterior Analytics* and the pseudo-Grossetestian Glosses to Boethius." In *The Scientific Works of Robert Grosseteste, Volume 1. Knowing and Speaking: Robert Grosseteste's De artibus liberalibus – 'On the Liberal Arts' and De generatione sonorum – 'On the Generation of Sounds'*, edited by Giles E. M. Gasper, Cecilia Panti, Tom McLeish, and Hannah E. Smithson, 367–383. Oxford: Oxford University Press, 2019.

Perler, Dominik, "Logik – eine 'wertlose Wissenschaft'? Zum Verhältnis von Logik und Theologie bei Roger Bacon." In *Logik und Theologie. Das Organon im arabischen und im lateinischen Mittelalter*, edited by Dominik Perler and Ulrich Rudolph, 375–399. Leiden: Brill, 2005.

Piché, David. *Le problème des universaux à la Faculté des arts de Paris entre 1230 et 1260*. Paris: Vrin, 2005.

Rosier-Catach, Irène. "Roger Bacon, al-Farabi et Augustin. Rhétorique, logique et philosophie morale." In *La Rhétorique d'Aristote. Traditions et commentaires de l'Antiquité au XVIIe siècle*, edited by Gilbert Dahan and Irène Rosier-Catach, 87–110. Paris: Vrin, 1998.

Wieland, Georg, "Ethik und Metaphysik. Bemerkungen zur Moralphilosophie Roger Bacon." In *Virtus politica: Festgabe zum 75. Geburtstag von Alfons Hufnagel*, edited by Joseph Möller and Helmut Kohlenberger, 147–173. Stuttgart-Bad Cannstatt: Frommann-Holzboog, 1974.

3 The Science of Weights

Jordan of Nemore's Influence on Roger Bacon's Scientific Method

Yael Kedar

3.1 Introduction

Roger Bacon is well-known for his introduction of the Islamicate science of *perspectiva* to the Latins, an enterprise that included an extensive and elaborate use of optical diagrams.[1] Bacon indeed placed much emphasis on geometrical demonstrations in optics, so much so that David Lindberg declares that "It was Bacon who showed Western scholars how to do optics in a mathematical way."[2] Lindberg also declares that Bacon was the first Latin writer to make extensive use of "serious and sustained geometrical argumentation."[3] Lindberg indicates three main sources for Bacon's extensive use of diagrams, which reflects his commitment to the search for geometrical causes in nature: Robert Grosseteste, Ptolemy, and Alhazen. While Grosseteste had mainly been a philosophical source of inspiration, in the works of Ptolemy and Alhazen, the promise of geometrical optics had been much more completely fulfilled.[4]

Bacon's conviction in the explanatory power of mathematics went, however, far beyond optics. He believed that "nothing can be known concerning the things of this world without the power of geometry."[5] Accordingly, he drew diagrams in order to solve problems of natural philosophy in domains that were far removed from optics. While these are not as numerous as the optical diagrams, they are of significance since Bacon thought of them as a model for the manner by which demonstrations in natural philosophy and metaphysics ought to be carried out in general. In the specific case which is the focus of this essay, Bacon's inspiration did not come from Ptolemy or Alhazen but from a lesser-known author, namely, Jordan of Nemore. In this chapter, I argue for the significance of Jordan for Bacon's scientific methodology, primarily concerning Bacon's call for the search for mathematical causes in nature. I suggest that while his geometry of optics was derived principally from Alhazen, Bacon's geometry of motion and his confidence in the efficacy of mathematics for the study of nature at large were derived from Jordan of Nemore. Hence, Jordan ought to be considered a significant source of Bacon's use of diagrams and, in general, of Bacon's scientific methodology. Moreover, Bacon's recognition of Jordan's achievement was unique in the thirteenth century. Even though his own mathematical capabilities were not impressive,[6] he was able to acknowledge the mathematical achievements of his time and integrate them

DOI: 10.4324/9781003405702-4

into his scientific outlook. He took both Alhazen and Jordan as models and inspiration for a new vision of how science ought to be performed.

I focus on one chapter from the fourth section of the *Opus maius*, titled "The Science of Weights," which has not been studied before, and analyse the chapter's proof that the arms of the scale will always return to balance if disturbed. I consider the problem that Bacon tried to solve, the method he used to solve it, and the context in which he discussed it. I draw conclusions as to Bacon's scientific method and his attempt to integrate the scientific achievements of his contemporaries into his vision of science. I conclude with a few remarks on Bacon's use of geometrical demonstrations in natural philosophy.

In the first section of this chapter, I present an outline of the state of the "science of weights" – a part of mechanics that we nowadays call "statics" – in the thirteenth century: its sources, methodology, and achievements. I then move on to discuss Bacon's take on the science of weights, which amounts to the proof that a disturbed scale carrying equal weights will return to balance. Lastly, I evaluate the influence of Jordan's corpus on Bacon's conception of the place of mathematics and geometry in the study of nature.

3.2 Jordan and the "Science of Weights" in the Thirteenth Century

The "Science of weights" is one of the seven special sciences which Bacon tried to promote in his campaign for the reorganisation of Christian studies at the universities. These included, besides the science of weights, perspective, astronomy, alchemy, agriculture, medicine, and experimental science. Bacon's inclusion of the science of weights in his classification was unusual; the science of weights did not appear in any other influential classification of his time. It appeared, though, in two earlier classifications, namely, the one proposed by Gundissalinus[7] and in al-Fārābī's *Catalogue of the Sciences*, written in the tenth century and translated to Latin in the twelfth century.[8] According to Gundissalinus, the science of weights was one of the seven mathematical sciences, along with arithmetic, music, geometry, optics, astrology, and engineering.[9] In the *Communia mathematica* Bacon listed these seven "practical and operative" sciences under practical geometry,[10] while in the beginning of the *Communia naturalium* he listed the same seven sciences as special natural sciences.[11] In both texts, he stressed their usefulness and utility, especially when it comes to the construction of instruments. The range of natural phenomena investigated by the science of weights is defined by Bacon thus:

> After the science of the heavens follows the science of the Elements in so far as they are the parts of the world, and this in so far as they are heavy and light, and therefore a science was established about them, which shows all the degrees and properties of heavy and light things, and of this kind is the science of Weights, in which heaviness and lightness and the motions of heavy and light things are certified, which Tebithus and Euclid, and many other authors, exemplify admirably.[12]

The inclusion of the science of weights by Bacon results from his acknowledgement of the new achievements in statics, made in the first half of the thirteenth century. Note, however, that he considered this science as including both statics and dynamics. It deals with all the motions of the elements, as derived from their tendencies to move towards the centre or the periphery of the sublunar world. It may be the case that the "science of motion" is a title that befits this science better.

The origins of statics date back to Archimedes' proof of the general law of the lever, which appealed to symmetry in the geometrical representation of the equilibrium of a lever with equal arms supporting equal weights. Static Archimedean demonstrations were wholly geometrical, inferring theorems from postulates in the Euclidian manner. An earlier contribution, characterised by a more dynamic approach, lacking the mathematical rigour of Archimedes, was expressed in a work titled *Mechanical Problems* attributed wrongly to Aristotle.[13] Another dynamic source was the ninth-century Arabic work composed by Thabit ibn Qurra called the *Liber karastonis*, which used the concept of virtual displacement in a germinal form.[14] Pseudo-Aristotle and Thabit alike understood the concept of virtual velocities (or displacement) in this way: the effective weight on a balance or lever increases proportionally with its distance from the fulcrum because the distance from the fulcrum endows it a greater velocity (because it traverses a larger arch in the same time). The effective weight is increased with velocity (namely, by the size of the arch).[15] To this was added a source of the static tradition, an anonymous Hellenistic work, *De canonio* (*On the Balance*), translated early in the thirteenth century.[16]

Later on in the thirteenth century, a new series of tracts on statical issues (*Scientia de ponderibus*) was composed, the most important of which were ascribed to Jordan of Nemore and his school. Jordan reworked the material of the Hellenistic mechanical traditions, and thus consolidated the union of Aristotelian dynamics with Archimedean mathematical statics.[17] Jordan developed three more pivotal statical principles besides the law of the lever, namely, the principle of the inclined plane, the principle of the composition of forces, and the principle of virtual displacement.[18] It is assumed, however, that Jordan was not himself the author of all the texts associated with his name. The statical texts ascribed to him were the following: (1) *Elementa Jordani super demonstrationem ponderum* (*The Elements of Jordan on the Demonstration of Weights*); (2) *Liber Jordani de Nemore de ratione ponderis* (*The Book of Jordan of Nemore on the Theory of Weights*); and (3) *Liber Jordani de ponderibus* (*The Book of Jordan on Weights*).

While there is an agreement in scholarly literature that the *Elementa* (1) is correctly ascribed to Jordan, and that *De ponderibus* (3) was an independent work,[19] a controversy evolved regarding the authorship of the *De ratione* (2). Duhem supposes that this work must have been written by a gifted disciple of Jordan, rather than by Jordan himself because it contains corrections of some of the errors of the *Elementa*.[20] Moody and Clagett, on the other hand, consider that Jordan himself may have come to recognise the errors of his earlier work, given that a few years elapsed between the composition of the *Elementa* and that of the *De ratione*.[21]

We know almost nothing about Jordan of Nemore. Neither his writings nor the surviving contemporary records yield any information about his life. All we know is that he taught in Paris[22] and was a capable mathematician, who was highly regarded by his contemporaries and successors, so much so that his name joined those of Euclid and Archimedes as an authority on the science of weights.[23] He composed treatises on algebra, arithmetic, geometry, and statics, all compiled before 1260.[24] Some of the texts ascribed to him were likely written by members of his circle, which included, among others, Richard of Fournival, Gerard of Brussels, and Campanus of Novara. This circle may have been active between 1241–1246, when Bacon was teaching in Paris. It has been suggested that Bacon and Jordan have met in the 1240s in Paris. On this suggestion, Bacon may have had close contact with Jordan and his circle, although he was not an actual member.[25]

The *Elementa* contains seven suppositions that form the basis for an original doctrine of "positional gravity" that opened a novel and promising way to compute the effectiveness of a weight on an incline. The *De ratione* expands the original nine theorems to 45 and adds two completely new sections. Brown describes it as "beyond question the finest work on medieval statics."[26] It contains the first correct solution to the inclined plane problem, corrections to errors in the *Elementa* regarding the bent lever and stable equilibrium, and it correctly uses the principles of work and static moment.[27] *De ponderibus* is important because it includes a prologue introducing the concept of "positional gravity" as a new notion and with full consciousness of the novelty of the term.

Positional gravity (*gravitas secundum situm*) is the component of a body's natural heaviness that governs its speed of descent on a constrained path.[28] According to this concept, a weight is heavier positionally, when, at a given position, its path of descent is less oblique. A more oblique descent is one in which, for a given distance, there is a smaller component of the vertical. This means that the force along an oblique path is inversely proportional to the obliqueness, where obliqueness is measured by the ratio of a given segment of the oblique path to the length of the vertical intercepted by that path. When the path is arched, the positional gravity or effective weight is measured by taking an arbitrary segment of the path of the possible descent or ascent of a weight and determining its projection on the vertical. The greater the vertical projection is in relation to the length of the trajectory, the heavier the positional gravity.

The concept of positional gravity is applied to problems of equilibrium in the following manner. If two weights, suspended on opposite ends of a balance beam are in equilibrium, this is because their positional gravities are equal. Their displacement will not cause their positional gravity to change, and if disturbed, they would return to a position of equilibrium. However, for two unequal weights, a displacement from the position of equilibrium would result in a gain of positional gravity for the weight elevated and equilibrium would be lost.[29]

The Jordanian texts address motion as a key to understanding heaviness and lightness. They acknowledge that motion can be simple or mixed.

A simple motion derives only from a body's natural inclination. In the case of mixed motion, an unnatural component (that is, the component of motion that does not follow the shortest path to the body's natural place) reduces a body's original heaviness while deflecting it from the line of natural descent. This interaction of natural and non-natural components determines the effective heaviness and the resulting velocity of a body.[30] Jordan had rewritten the conditions for geometrising constrained motion. To compare motions, his demonstrations included diagrams in which the vertical lines represented the simple motion of free fall and a non-vertical line represented compound motion. A vertical line indicated that the body traversing it possessed its full heaviness, while the non-vertical lines – sometimes oblique and at other times curved – showed the descending body to have less heaviness than in free fall.[31] This geometrisation leads us back to Bacon and his scientific methodology.

3.3 Bacon's "Science of Weights"

Bacon devoted one short chapter of the *Opus maius* to the science of weights (Part IV, Distinction 4, Chapter 16), almost all of which revolves around one single proof, namely, proposition 2 from *De ponderibus*,[32] which appears in a different and more geometrically elaborated form also in *De ratione* (theorem R1.02).[33] This short chapter on the science of weights indicates that Bacon was not interested in the science of weights per se, but rather in its method. He declared that he wanted to disclose "the potency of geometry in motions," namely, to demonstrate how effective geometry is for the study of motion. It is one of three natural philosophy demonstrations, all found in the fourth distinction of Part IV of the *Opus maius*, framed by Bacon as *propter quid*, and supposed to provide the geometrical reasons or causes of natural phenomena.[34] Two references to Jordan from the *Communia mathematica* provide a strong indication for this contention. The first reference sets an opposition between books that present speculative geometry "by way of narration, which is *quia*" (*per viam narracionis et quia est*), and books such as the *Libro elementorum Euclidis*, and *Liber Jordani de triangulis*, providing explanation "by reason and *propter quid*" (*per causam et propter quid*).[35] The second reference is about speculative arithmetic, handed down by Euclid and Boethius "by way of cause" (*per viam cause et propter quod*), even though Boethius "does not seem to use demonstrations in the same way as Jordan" (*non videatur sic demonstracionibus uti sicut Jordanus*). Although the virtue of Jordan's demonstrations arises from Boethius, "few [things] must be demonstrated of numbers under the method of demonstration which Jordan uses" (*pauca de numeris demonstranda sunt sub modo demonstrandi quo utitur Jordanus*).[36] Now, which method is that? Concisely put, what Bacon is advocating is a method of proof that seeks to find geometrical causes for motion and change of bodies in the sublunar world. Bacon thought that the best way to find these geometrical causes is to use diagrams that visually present the relations between geometrical objects such as lines, arches, and angles, and thus address the causes of natural phenomena, rather than their mere descriptions. As Fisher and Unguru note, for Bacon mathematics offers the ground for certainty by providing perceptible figures for all its results: "mathematics can provide an example sensible to all, and experience perceptible to the senses through the drawing of figures and counting, so that all is made clear to the senses."[37] The geometrical diagrams and mathematical counting furnish sensible experiences that validate its proofs. When there are doubts and disagreements, the drawing of figures will settle the issue:

> in the other sciences, there are doubts, opinions, and contradictions on our part, so that there is hardly agreement on the smallest question, much less on a sophism; for in those sciences, by their nature, there is nothing in them to govern the drawing of figures and [the determination of] computations, by which all things must be proven true. For that reason, only in mathematics does certainty exist without doubt.[38]

Bacon was a harsh critic of the Scholastic method of commenting on Aristotle and Lombard's *Sentences*. He had a much lower opinion of logic as a science than most of his contemporaries.[39] He argued that logical arguments alone, even when they originate from experience, are not sufficient for verification; mathematics, by contrast, is the only means of arriving at a true demonstration.[40] He put forward the strong claim that mathematics can vouch for the truthfulness of the premises of all knowledge. He held that our intellect is best suited to grasp quantities,[41] and therefore our logical processes are founded in quantities. Accordingly, our logical categories, as portrayed by Aristotle, depend on mathematics: "all of the categories depend on an understanding of quantity, which is the province of mathematics; and therefore, the whole value of logic depends on mathematics."[42]

Moreover, mathematical demonstrations are characterised by an order and clarity that ought to be transferred to the other sciences. Hence, the mathematical method should serve as a model for demonstrations in the other sciences, and even be implemented within them:

> It is therefore shown that mathematics is wholly necessary and useful to the other sciences. The reasons for this are general, but particular among them are those showing how the application of mathematics penetrates all parts of philosophy, resulting in an understanding of all things. And this shows nothing other than that the other sciences do not owe their knowledge to dialectical and sophistical arguments, presented jointly, but through mathematical demonstrations that penetrate to truths and to the works of the other sciences, thereby regulating them; without them, they can neither be understood, demonstrated, taught, or learned.[43]

Bacon's call to establish the mathematical method as general and overarching for all the sciences was meant to replace the method of "dialectical and sophistical arguments" which results in insoluble disagreements. He wanted to use the mathematical method to resolve the debates and settle the conflicting positions that render the sciences dubious. The way to achieve this, within the frame of the *Posterior Analytics* scientific methodology, is to use mathematical theorems in demonstrations of the other science.

In the chapter on the science of weights, Bacon tried to put these general methodological declarations into practice. He started out from theorem R01.2 (proposition 2, *De ponderibus*): two beams of equal length, carrying equal weight will be in equilibrium, and if disturbed, they will return to a horizontal position. He defined the centre of revolution of the arms of the scale as the centre of an imaginary circle formed by the descending and elevating movements of these equal arms.[44] To make the concept of the centre of revolution clearer, he drew a Figure 3.1. The figure is very similar to those drawn in the *Elementa* and *De ratione ponderibus*.[45] The centre of revolution is *d*, and *ab* – the arms of the scale.

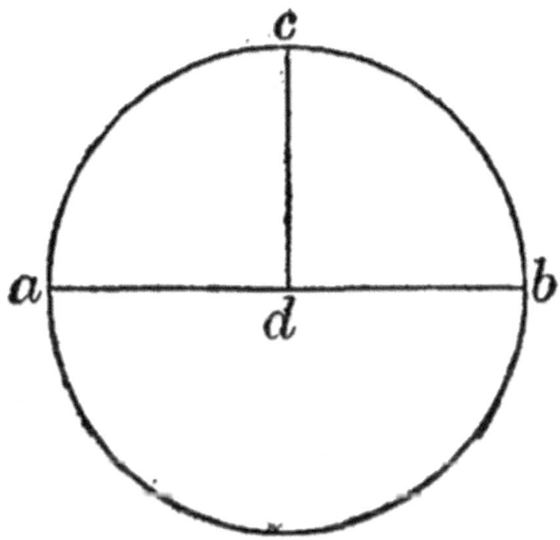

FIG. 19.

Figure 3.1 Roger Bacon, *Opus maius*.

He then moved on to the problem that intrigued him. *De ratione* argues for the acceleration of falling bodies: "The longer a heavy body falls, the faster it becomes in descending" (R4.06).[46] The cause of acceleration is the medium: the heavy body pushes the parts of the medium beneath it, and they move the parts beneath them. In this way, the more the body advances, it encounters less resistance and thus becomes heavier, thus pushing the receding parts even more. "Hence its velocity also observed to be continuously multiplied" (R4.06).[47]

Although this rule is taken from *De ratione*, Bacon provided a reference to Aristotle's *De caelo* as the source. The referred text, however, (*De caelo* 4.4) makes a distinction between absolute and relative heaviness and does not address the speed of falling bodies. As Duhem rightly notes, Aristotle and his commentators (including Alexander of Aphrodisias, Themistius, Simplicius, Averroes, and Thomas Aquinas) have written about the function of the medium in the **continuation** of a body's motion, not about its function in a body's acceleration.[48]

Bacon accepted wholeheartedly that the velocity – and hence also the weight – of a falling body increases as it descends. In the *Communia naturalium,* he provided a different explanation for this rule, by which the force (*virtus*) of the natural place is added to the force (*virtus*) internal to the falling body.[49] He saw, however, a problem settling this rule with theorem R01.2. For if one would depress one arm of a balanced scale, namely, one in which both arms are of the same length and carry the same weight, then the lower arm will become heavier, the higher arm will become lighter, balance will be lost, and the two arms will never return to a horizontal position. Moreover, this contradicts Jordan's concept of positional gravity and his claim that the lower arm is lighter positionally. Of course, this is also contrary to experience since we can see that the beam oscillates and returns to balance. Bacon seems to be missing here the recognition of the difference between statics and dynamics or between constrained and non-constrained motions.

Nevertheless, Bacon declared he would solve this apparent difficulty. He drew a circle (Figure 3.2) representing the scale's arms' possible path around the centre of revolution and argued that the drawing would address objections because the intellect "can only obtain clarity through a figure":[50]

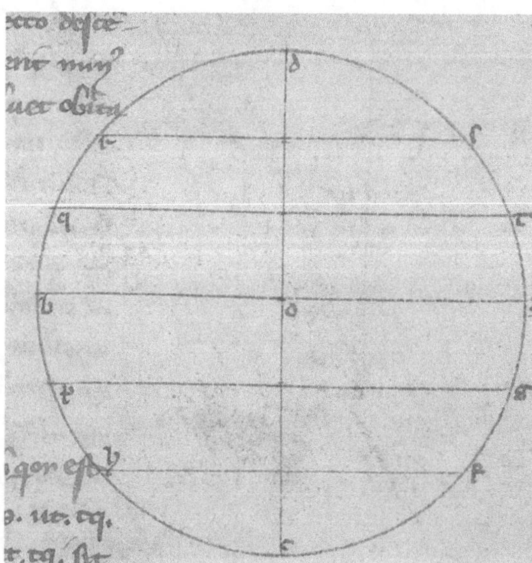

Figure 3.2 Roger Bacon, *Opus maius*, MS Vat. Lat. 4086, fol. 28r, Biblioteca Apostolica Vaticana, Vatican City.

O is the centre of revolution, *ab* the horizontal diameter, and *dc* the vertical. The circle is cut into equal arches, and lines parallel to the horizontal diameter are drawn between the ends of the equal arches (*fb*, *gp*, *tq*, *sr*). These lines cut unequal segments of the vertical (*dc*): the parallels closest to *ab* on both sides (for example *tq*) will cut a larger segment of *dc* than the more distant parallels (such as *sr*).

Now, say the scale is moved and one arm is elevated to *t* while the other descends to *p*, thus forming the line *tp*. The scale will be unbalanced. If the lower arm keeps descending all the way from *p* to *h*, its path will cut a smaller segment of the vertical *dc* than the upper arm descending from *t* to *a*, since the segments closer to the centre are larger than those further away. Thus, the upper arm will be heavier positionally in relation to the lower arm.

Bacon added the following Figure 3.3, which is almost identical to the *De ratione* figure:[51]

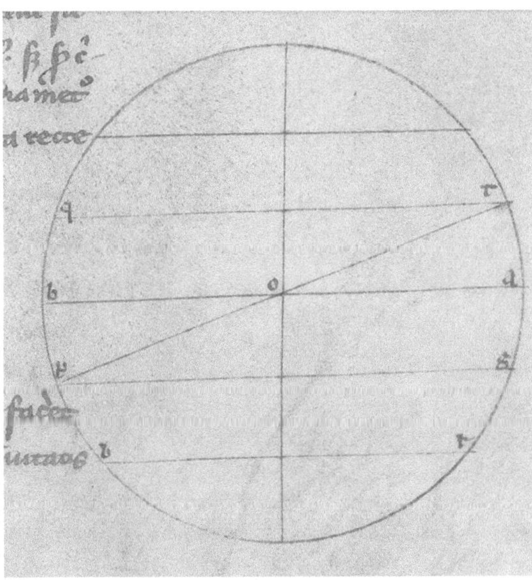

Figure 3.3 Roger Bacon, *Opus maius*, MS Vat. Lat. 4086, fol. 33r, Biblioteca Apostolica Vaticana, Vatican City.

Supposedly, since the descending arm *p* will be closer to the centre of Earth, it would be heavier according to the rule that the weight gets heavier the closer it gets to the centre of Earth. Arm *p*, however, does not move in a straight path towards the centre, but along the circle's circumference, and its course moves less directly than that of arm *t* along the diameter drawn to the centre of Earth. This renders *t* heavier positionally than *p*. It will therefore descend, pulling *p* upwards. Bacon drew the following diagram (Figure 3.4), comparing the paths of *t* and *p* to the centre of Earth. This diagram is not found in *De ratione*:[52]

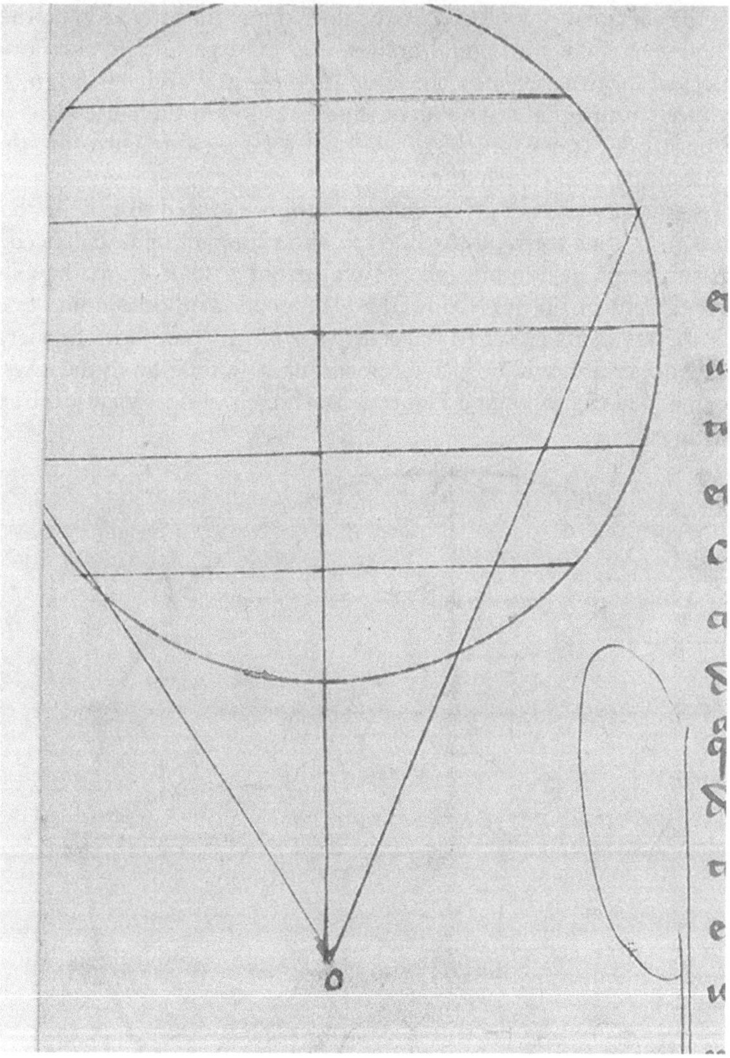

Figure 3.4 Roger Bacon, *Opus maius*, MS Vat. Lat. 4086, fol. 34r, Biblioteca Apostolica Vaticana, Vatican City.

Due to the rule of acceleration, *p* should have been heavier since it is closer to the centre of Earth. But its trajectory is more oblique than that of arm *t*, and this renders *t* heavier. The weight added to *p* by its proximity to the centre of Earth becomes insignificant, Bacon argued. It is too small to have an effect.

The whole demonstration is based on the postulates given at the beginning of *De ratione* and at the beginning of *De ponderibus*:[53] a body is heavier in descent the more its motion towards the centre of the Earth is direct

(RI.003), with the obliqueness of descent and positional heaviness being inversely related (RI.004). That is, a heavier body will have a less oblique path. Its obliqueness is determined by the segment of the vertical it cuts: the greater the segment in relation to its path, the greater its obliqueness (RI.005). Lastly, Bacon relied on RI.006 when he argued that the descent of *t* renders *p* positionally lighter. Bacon referred to the last postulate explicitly: "Certainly, when Jordan says in his third proposition that the higher arm is less heavy because it follows the descent of the other arm with an opposite motion, namely an ascent, I gladly concede it."[54]

He added that the continuous motion of the scale, which oscillates before its arms reach the final equilibrium, is not caused by the nature of the weights but by the impact of the particles of the air which hold the motion imparted to them by the scale, vibrating for a while.[55]

Although, as we can see, Bacon had some success in presenting Jordan's geometrical reasoning, it seems as though he did not properly realise the difference between a constraint, static system and a dynamic one. He mixed geometrical and physical considerations, favouring the geometrical arguments, but without completely dismissing the physical causes. It can be rightly argued that Bacon did not contribute to the theory of weights. He added nothing of value to the Jordanian contribution. The significance of his thought is found elsewhere, though, in the methodological domain. Despite the shortcomings of his grasp of the theory proposed by Jordan, one can acknowledge that Bacon was able to identify and appreciate Jordan's achievements. This cannot be said about his more well-known contemporaries, such as Albert the Great and Thomas Aquinas. Bacon had contacts with working mathematicians of his time, and he was eager to promote their cause and to integrate their findings into his own synthesis. His belief that science is a joint human enterprise did not remain at the level of empty declarations; he practised it. I contend that Bacon considered the Jordanian corpus a model of how science ought to be carried out, and indeed made attempts to apply this model to domains other than statics.

3.4 Bacon's Application of the Jordanian Methodology

I have found at least two cases in which Bacon tried to apply what he learnt from Jordan. The first case concerns the supposed heat caused by a body moving downward to its natural place, and the second concerns the contraction of water the closer it gets to the centre of Earth. The two cases are different, but both involve the concepts of heaviness, natural place, and geometrical causes for the physical phenomena.

The concept of "positional gravity" exemplified the idea of the resolution of a force into various components. Jordan showed that the motive power by which the body was moved could be disassociated into two: the natural gravity downward towards the centre, and a violent horizontal force, constraining the weight to move along the curve instead of descending vertically.[56]

Bacon adopted in full the idea of motion as resulting from a composition of forces. In the chapter preceding the one on the motion of the scale, he demonstrated geometrically that the natural motion of a body downward involves strain and, therefore, heat. He claimed that the use of a geometrical figure will prove that natural philosophers were wrong in arguing that the natural motion of bodies involves no strain.[57] He drew (Figure 3.5) a piece of wood in the air (abc), the centre of which is in a straight line above the centre of the Earth. The middle and two extreme points of the wood fall along parallel lines, only one of which goes directly to the centre of the Earth. Any deviation from the centre, Bacon argued, involves a non-natural or violent force. Hence, only the middle point of the wood falls along a straight path naturally, with no added violent force. There is more than one force acting on a heavy body moving to its natural place, pulling it downwards and in opposite directions. One force attracts the middle point along a straight line, and the other forces draw the external points in other directions, along oblique paths. Because they fall in different directions, the parts of the body are pushed away from one another and thus the body rarifies, and rarefaction generates heat. Heat is the effect caused by the operation of opposing forces.[58]

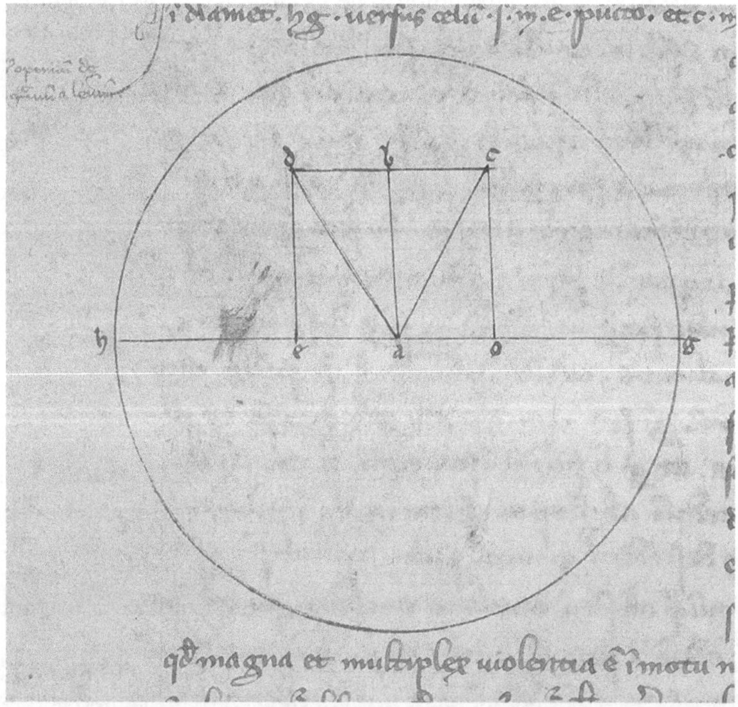

Figure 3.5 Roger Bacon, *Opus maius*, MS Vat. Lat. 4086, fol. 34v, Biblioteca Apostolica Vaticana, Vatican City.

In the *Liber de triangulis Iordani* ascribed to Jordan, the ratio of arches and plane segments in different circles is addressed. Bacon relied on theorem 28 from this book[59] for his proof that water contracts as it gets closer to the centre of Earth. He argued that if we were to place two cups, one at a higher place, such as a balcony, and another at a lower place, such as a cellar, and then fill them with water, the "[water will] form itself into a smaller and larger sphere with respect to the center of the Earth."[60] The water's surface of the higher cup will be a portion of a sphere more distant from the centre of the Earth and therefore larger than the surface of the water in the lower cup. The diameter associated with each cup will then cut an equal chord (assuming the cups are equal) of each sphere; however, the proportions between the chords and the spheres will be different: the chord of the higher cup will form a smaller part relative to the larger sphere, while the chord of the lower cup will be relatively larger. Consequently, the water's surface of the two cups will differ: the smaller sphere of the lower cup will result in a higher surface, while the higher cup's surface will be lower. Given this situation, Bacon inferred that the lower cup can contain more water than the higher one since the same amount of water will require less space or, in other words, the water in the lower cup will contract. Bacon wrote:

> By proposition XXXVIII in Jordan's *De triangulis*, in circles that are not equal, the same chord cuts a smaller portion from the larger circle and a larger portion from the smaller circle, and thus it would also be for spheres.[61]

Here, once again, Bacon relied on principles he found in Jordan and used them to produce new knowledge in natural philosophy. In both cases, he used the relations between geometrical objects to account for the behaviour and properties of bodies in motion.

3.5 Conclusions

It is obvious that Bacon placed much emphasis on geometrical demonstrations in optics. We know that his source when it comes to this discipline was Alhazen. The significance he ascribed to geometrical demonstrations of motion is less known. I suggest that while his geometry of optics was derived from Alhazen, his geometry of motion followed that of Jordan. I concede that Bacon did not seem to have an in-depth understanding of the mathematical and statical principles developed by the Jordanian school. He had a much better grasp of the optical principles he found in Alhazen. He did, however, have a good appreciation of the school's achievements, and he thought that these achievements had to do with the use of geometry to find natural causes. This was the common feature linking the science of *perspectiva* with the science of weights, and Bacon thought that the same method should be extended to the other sciences, not only to the seven special ones he promoted but also to the other disciplines as well. The influence of Jordan was significant in this respect because it demonstrated that geometry can be a productive tool not only in the study of light and vision but also in the investigation of the motions of bodies and the effects of this motion.

Notes

1 For more on Bacon and his debt to Islamic optics, see Nader El-Bizri's chapter in this volume.
2 Lindberg, "Roger Bacon and the origins of *perspectiva* in the West," 264.
3 Lindberg, "Roger Bacon and the origins of *perspectiva* in the West," 259.
4 Lindberg, "Roger Bacon and the origins of *perspectiva* in the West," 258.
5 Bacon, *Opus majus* IV.4.23, 249. Unless stated otherwise, all the quotations from the *Opus majus*, Part IV are taken from the translation by Dennis, *Roger Bacon's Mathematical Thought*. The page numbers refer accordingly to this translation by Dennis. The Latin text of the *Opus maius* can be found in the three-volume edition by Bridges.
6 See Molland, "Roger Bacon's Knowledge of Mathematics," 157.
7 Some of the most influential classifications are presented and discussed in Hackett, "Roger Bacon on the Classification of the Sciences."
8 Brown, "The Science of Weights," 188.
9 See Weisheipl, "Classification of the Sciences in Medieval Thought," 219.
10 Bacon, *Communia mathematica*, I.3.3, 44–45: "SECUNDA pars principalis Geometrie practice est de composicione instrumentorum omnium scienciarum, et hec habet septem partes principales secundum numerum septem scienciarum particularium practicarum et operativarum in corporibus .... Quarta pars docet formare omnia instrumenta sciencie ponderum, in quibus est magna utilitas et multis modis."
11 Bacon, *Communia naturalium*, I.1.1.2, 5: "Declaravi igitur in illa parte Methamatice quod preter scienciam communem naturalibus sunt septem speciales: videlicet, Perspectiva: Astronomia, judiciaria et operativa: Sciencia ponderum de gravibus et levibus: Alkimia: Agricultura: Medicina: Sciencia Experimentalis."
12 Bacon, *Communia naturalium*, I.1.1.2, 6: "Post scienciam celestium sequitur sciencia de Elementis in quantum sunt partes mundi, et hoc in quantum sunt gravia et levia, et ideo constituta est sciencia de eis, que omnes gradus et proprietates gravium et levium ostendit, et hujusmodi est sciencia de Ponderibus, in quibus certificantur gravitas et levitas et motus gravium et levium, que apud Tebith et Euclidem et multos alios auctores egregie ventilatur."
13 Moody and Clagett, *The Medieval Science of Weights*, 4. For the main ideas of this text, see Brown, "The Science of Weights," 180–183.
14 For a description of this work, see Brown, "The Science of Weights," 187.
15 Moody and Clagett, *The Medieval Science of Weights*, 5.
16 Brown, "The Science of Weights," 188.
17 Moody and Clagett, *The Medieval Science of Weights*, 4.

18 Moody and Clagett, *The Medieval Science of Weights*, 15.
19 Moody and Clagett, *The Medieval Science of Weights*, 9 and 145.
20 Duhem, *The Origins of Statics*, 99–100.
21 Moody and Clagett, *The Medieval Science of Weights*, 18 and 171.
22 Hoyrup, "Jordanus de Nemore, 13th Century Mathematical Innovator," 346.
23 Brown, "The Science of Weights," 191.
24 Moody and Clagett, *The Medieval Science of Weights*, 6–7.
25 Hoyrup, "Jordanus de Nemore, 13th Century Mathematical Innovator," 345–351.
26 Brown, "The Science of Weights," 197.
27 Brown, "The Science of Weights," 197.
28 Brown, "The Science of Weights," 191.
29 Moody and Clagett, *The Medieval Science of Weights*, 16.
30 Brown, "The Science of Weights," 196.
31 Brown, "The Science of Weights," 191.
32 The proposition is found in Moody and Clagett, *The Medieval Science of Weights*, 157.
33 Found in Moody and Clagett, *The Medieval Science of Weights*, 177–179.
34 For an extended analysis of these demonstrations as *propter quid*, see Kedar, "Propter quid Demonstrations."
35 Bacon, *Communia mathematica*, I.3.2, 42.
36 Bacon, *Communia mathematica*, I.3.3, 47.
37 Bacon, *Opus majus* IV, 1.3, 74.
38 Bacon, *Opus majus* IV, 1.3, 75.
39 Fisher and Unguru, "Experimental Science and Mathematics," 373.
40 Fisher and Unguru, "Experimental Science and Mathematics," 373.
41 See Demange, "Cupiens Mathematicam Tractare."
42 Bacon, *Opus majus* IV, 1,2, 68–70. See also Bacon, *Communia mathematica*, 13.
43 Bacon, *Opus majus* IV, 1.3, 78.
44 Bacon, *Opus majus* IV, 4.15, 175–176.
45 The diagram is taken from Bridges' edition of the *Opus majus*, vol.1, 170.
46 Moody and Clagett, *The Medieval Science of Weights*, 217.
47 Moody and Clagett, *The Medieval Science of Weights*, 217.
48 Duhem, *The Origins of Statics – the Sources of Physical Theory*, 101.
49 See Kedar, "Virtus and Species."
50 Bacon, *Opus majus* IV.4.15, 178. The diagram is taken from Vat. Lat. 4086, fol. 34r. This is a fourteenth-century manuscript, closely resembling (but not a copy of) Cotton Jul. D. v, which is the oldest manuscript of the *Opus majus*, held by the British Museum and partly destroyed by fire. See Little, "Appendix: Roger Bacon's Works," 379.
51 This diagram is from Vat. Lat. 4086, fol. 34r. The *De ratione* figure is R.02b, in Moody and Clagett, *The Medieval Science of Weights*, 178.
52 Vat. Lat. 4086, fol. 34v.
53 Moody and Clagett, *The Medieval Science of Weights*, 155 and 175.
54 Bacon, *Opus majus* IV, 4.15, 182.
55 Bacon, *Opus majus* IV, 4.15, 182.
56 The idea of the composition of forces can be found in Aristotle, long before Jordan (see Duhem, *The Origins of Statics – the Sources of Physical Theory*, 81–83). Jordan, however, interpreted this idea aided by the geometrical tradition of Euclid and Archimedes, and Bacon tried to implement it according to his model.
57 Bacon, *Opus majus* IV, 4.14, 171.
58 The diagram is taken from Vat. Lat. 4086, fol. 33r.

59 Theorem 28 appears, in fact, in *Liber philotegni*, a shorter version of *Liber de triangulis Iordani.*
60 Bacon, *Opus majus* IV, 4.11, 159.
61 Bacon, *Opus majus* IV, 4.11, 157.

Bibliography

Primary Sources

Roger Bacon. *Communia mathematica.* Edited by Robert Steele. *Opera hactenus inedita Rogeri Baconi*, fasc 16. Oxford: Clarendon, 1905–1940.
Roger Bacon. *Communia naturalium.* Edited by Robert Steele. *Opera hactenus inedita Rogeri Baconi*, fascs 2–4. Oxford: Clarendon, 1905–1940.
Roger Bacon. *Opus majus.* Edited by John Henry Bridges. *3 vols.* Oxford and Edinburgh: Clarendon, 1897–1900.
Roger Bacon. "Opus majus IV." In *Roger Bacon's* Mathematical Thought: A Translation *of Part IV of the Opus majus with Introduction and Commentary*, translated by Patrick Willard Dennis. Dallas: Unpublished Ph.D dissertation, 2011.

Secondary Sources

Brown, Joseph E. "The Science of Weights." In *Science in the Middle Ages*, edited by David C. Lindberg, 179–204. Chicago: University of Chicago Press, 1978.
Demange, Dominique. "'...cupiens mathematicam tractare infra radices metaphysice...'. Roger Bacon on mathematical abstraction." *Revista Española de Filosofía Medieval* 28/1 (2021): 67–98.
Duhem, Pierre, *The Origins of Statics – the Sources of Physical Theory*. Translated by Grant F. Leneaux, Victor N. Vagliente, and H, Guy. Dordrecht: Kluwer Academic Publishers, 1991 [1905–6].
Fisher, N. W. and Unguru, Sabetai. "Experimental Science and Mathematics in Roger Bacon's Thought." *Traditio* 27 (1971): 353–378.
Hackett, Jeremiah. "Roger Bacon on the Classification of the Sciences." In *Roger Bacon and the Sciences – Commemorative Essays,* edited by Jeremiah Hackett, 49–66. Brill: Leiden, 1997.
Hoyrup, Jens. "Jordanus de Nemore, 13th Century Mathematical Innovator." *Archive for History of Exact Science* 38 (1988): 307–363.
Kedar, Yael. "Virtus and Species in the Philosophy of Nature of Roger Bacon." *Revista Española de Filosofía Medieval* 28/1 (2021): 43–66.
Kedar, Yael. "Propter quid Demonstrations: Roger Bacon on Geometrical Causes in Natural Philosophy." *Synthese* 203/1 (2024): 1–24.
Lindberg, David C. "Roger Bacon and the Origins of *Perspectiva* in the West." In *Mathematics and its Applications to Science and Natural Philosophy in the Middle Ages,* edited by Edward Grant and John Emery Murdoch, 249–268. London: Cambridge University Press, 1987.
Little, A. G. "Appendix: Roger Bacon's Works, with References to the MSS. With Printed Editions." In *Roger Bacon – Essays*, edited by A. G. Little, 373–425. Oxford: Clarendon Press, 1914.
Molland, George. "Roger Bacon's Knowledge of Mathematics." In *Roger Bacon and the Sciences: Commemorative Essays*, edited by Jeremiah Hackett, 151–174. Leiden: Brill, 1997.

Moody, Ernest A. and Marshall Clagett, eds. *The Medieval Science of Weights (Scientia de ponderibus), Treatises Ascribed to Euclid, Archimedes, Thabit ibn Qurra, Jordanus de Nemore, and Blasius of Parma*. Madison: The University of Wisconsin Press, 1952.

Weisheipl, James A. "Classification of the Sciences in Medieval Thought." *Mediaeval Studies* 27 (1965): 54–90.

4 Roger Bacon and John Pecham on Arithmetic

Aurélien Robert

In his ambitious project to reform education in Europe, Roger Bacon gave a prominent place to the mathematical sciences. Indeed, in the treatises he sent to Pope Clement IV, he lamented the lack of mathematical knowledge among the intellectuals of his time: only two were excellent (*perfecti*): John of Tynemouth and Peter of Maricourt, while two others, Campanus de Novara and a certain Master Nicolas, tutor to Amaury of Montfort, were only good (*boni*).[1] Of course, he mentions his contemporary fellows Robert Grosseteste and Adam Marsh as eminent figures who were aware of the power of mathematics to explain natural – and also supernatural – phenomena.[2] But Roger Bacon wanted to give a more complete account of the nature and role of mathematics in relation to the other sciences. For in his view, mathematics is useful for all the sciences, whether speculative or practical, directed towards human affairs or towards God.

Within this programme, which he developed not only in the *Opera* he sent to the Pope but also in some of his philosophical works, above all the *Communia mathematica*, one finds precise descriptions of the traditional sciences of the quadrivium (arithmetic, geometry, astronomy, and music).[3] While his treatment of geometry and its application to the field of optics has been studied in great detail,[4] as well as his discussions of astronomy[5] and music,[6] arithmetic has been somehow neglected. The aim of this chapter is therefore to examine the role and place of arithmetic in Roger Bacon's system. For if arithmetic has often been considered, since Boethius, as the first and most important science of the quadrivium, this does not seem to be the case with Roger Bacon. We shall therefore try to understand the reasons for this apparent depreciation of arithmetic in favour of geometry and, above all, *perspectiva*.

To highlight the specificity of Roger Bacon's analysis, we will compare his views with those of John Pecham, another thirteenth-century Franciscan whose philosophy is often regarded as very close to the Baconian synthesis.[7] Indeed, it is often said that there was a shared taste for mathematics within the Franciscan Order, especially after Robert Grosseteste taught in some of their *studia*. But the reality is that behind this common interest, the place given to the different sciences of the quadrivium varies, thereby leading to different philosophical approaches to mathematics.[8] This appears quite

DOI: 10.4324/9781003405702-5

clearly when one compares Roger Bacon and John Pecham's respective positions on the division and composition of the continuum.

The philosophical and mathematical question of the continuum reveals the way in which they considered the relationship between discrete and continuous quantities, that is to say, between arithmetic and geometry. If Roger Bacon refuses to give primacy to numbers over continuous quantities, John Pecham remains more faithful to a long tradition of Pythagorean and Platonic origin, which places arithmetic at the centre of mathematical sciences, to the point of making it the gateway to all knowledge of the created and uncreated world. The main consequence of this philosophy of mathematics is that Roger Bacon rejects all forms of atomism, including mathematical indivisibilism, whereas John Pecham seems more sympathetic with this theory.

4.1 Mathematics in Roger Bacon's Classification of the Sciences

Roger Bacon's considerations on mathematics in general and arithmetic, in particular, cannot be understood independently of the general context in which they occur, i.e., a general defence of Christendom.[9] In his view, the promotion of his model for organising knowledge would ensure the cultural superiority of Christian ideals, reinforce their moral values, and even persuade and convert the infidels. This is based on the assumption that even the most speculative sciences also have a practical aspect and must be cultivated with a moral aim.[10]

In the *Opus maius*, when the study of grammar, languages, and mathematics is presented as the most necessary and useful skill for acquiring all kinds of knowledge, these sciences are by no means presented as the highest or noblest sciences. They are always subordinated to moral philosophy, understood here as Christian theology.[11] This explains why a very large part of Roger Bacon's discourse on mathematics is focused on its *utilitas* or *potestas*, i.e., its possible application in different fields of philosophy and theology. Indeed, as George Molland rightly pointed out, Roger Bacon never really wrote a mathematical treatise, but mostly dealt with the usefulness and applicability of mathematics.[12]

In this respect, mathematics is not only related to theology and moral philosophy but also to natural philosophy and metaphysics. For Roger Bacon, it goes without saying that the four sciences of the quadrivium are oriented towards the higher science of *perspectiva*, which is supposed to explain the whole structure of the world (*totam machinam mundi*), and also to what he calls *scientia experimentalis*.[13] As he states very clearly in his *Opus tertium*:

> It is therefore not surprising that everything is known through mathematics, because all sciences are interconnected (as I said above), although each of them also has its own characteristics. Therefore, each science has some power (*potestatem*) on another, and one cannot know one science without all the others, as has already been said. Thus it is

> necessary that everything be known through this science, for all the actions of things is done by the multiplication of species and forces which proceed from the agents in this world in material patients.[14]

While the natural end of mathematics, and more specifically geometry, is the application of its laws to nature and natural causality within the science of *perspectiva*, Roger Bacon insists that mathematics is also present in all kinds of knowledge. Moreover, even when he defends a particular order in the teaching and learning of these sciences, the fact remains that they are all connected and mutually enriching. Thus, the idea of a strict hierarchy of sciences should not be taken literally, because they all depend on each other. They do not only share the same type of method as "sciences," they are also complementary for the understanding of their respective objects.

As Jeremiah Hackett has shown, Roger Bacon's classification of the sciences is based on a multiplex heritage.[15] It includes elements of the Boethian distinction between speculative and practical sciences, Hugh of Saint Victor's distinction between the human and divine sciences (without the addition of the mechanical sciences as a separate branch of knowledge), and the influence of the Aristotelian corpus, in particular the *Posterior Analytics*, with the mediation of Arabic thinkers such as al-Fārābī (and its adaptation by Gundissalinus), Avicenna, and Averroes.[16] The result is a classification of the sciences that is original in several respects.

In the *Communia naturalium*, for instance, the organisation of the sciences is presented in an order very similar to that of the chapters of *Opus maius*: grammar, logic, mathematics, natural philosophy, metaphysics, and moral philosophy.[17] Each of them is divided into general and special sciences. Hackett has studied very closely the subdivisions of natural philosophy as presented in the *Communia naturalium*, which are of particular interest because Roger Bacon distinguishes seven special natural sciences, which seems quite unusual. These are, in order: *perspectiva*, astronomy, the science of weights, agriculture, medicine, alchemy, and experimental science.

It may seem surprising to find *perspectiva* and astronomy among natural sciences, as they seem to belong to mathematics. But this can be understood as soon as one realises that each science has one or more theoretical parts and one or more practical parts. It may be the case that within the same science, or at least within a science called by a single name, one part belongs to mathematics and another to natural philosophy. This is the case with *perspectiva*, which has a speculative part based on geometry and a practical part that seeks to apply these laws to natural phenomena and to optics in particular. This is also the case with astronomy, which has a mathematical part and at least two practical parts: one concerning the use of tables and instruments and the other the use of astronomical data for practical purposes (it is then called astrology).[18] Of course, mathematics is also present in the science of weights, but also, as we shall see, in agriculture, and most plausibly in medicine and experimental science.

In this classification of the sciences, mathematics is not presented as the very first science. In fact, it appears after grammar and logic. But here again, we must be wary of too strict an order. Of course, one needs the rudiments of grammar to formulate mathematical propositions, as well as elements of logic to construct valid demonstrations. But according to Roger Bacon, the grammarian cannot perfect his science without mathematics, particularly when it comes to rhythm and, more generally, phonetic quantities.[19] In the same way, logic – which here includes rhetoric and poetics, according to the extended vision of the *Organon* defended by the Arabic tradition – cannot be perfected without mathematics. From the elementary level of Aristotle's *Categories*, Roger Bacon explains, the logician cannot understand the category of quantity without mathematics.[20] Similarly, how can the logician understand the superiority of mathematical reasoning in the *Posterior Analytics* without being a mathematician himself?[21] The same applies to rhetoric and poetics, which depend partly on music, understood as a part of the quadrivium.

If Roger Bacon continues to give mathematics a certain primacy, despite the fact that this science, like any other, requires some skills in grammar and logic, it is partly because this has been a traditional way of presenting this science since Boethius, but also, and above all, because he sees mathematical reasoning as almost innate (*quasi innata*).[22]

4.2 Mathematics as an Innate Form of Logical Reasoning

On the first point, he quotes a number of *auctoritates*, among which Boethius is probably the most important:

> Regarding authority I thus begin. In the second prologue to his *Arithmetica*, Boethius says that "if an inquirer is not familiar with these four sciences, he is not able to find the truth." He continues, "No truth is correctly discerned without this kind of speculation." To this end, Boethius says that "He who spurns these sciences, the paths of wisdom, does not philosophise well." He continues, "Whoever overlooks this allows the destruction of the knowledge of all learning." What is more, Boethius confirms the judgement of all important men, saying that "Among all men of ancient authority, who prevailed with a purer understanding under the leadership of Pythagoras, it is clear that no one achieves perfection in the discipline of philosophy, unless the nobility of such wisdom is examined within the so-called quadrivium."[23]

In another passage, citing Socrates' philosophy of mathematics through Cicero, he proves that we can discover mathematical truths all by ourselves, before any formalised teaching.[24] What is more, mathematics would be the oldest science, a sign that it is also the most natural for the human mind. It is also the easiest science, since lay people and even illiterate people can draw figures, count, or sing. Finally, it is the most certain, because its demonstrations, based on necessary causes, arrive at necessary truths.[25] As a consequence, if the other sciences were based on mathematics, they would reduce their uncertainty.

It is therefore important to understand that the primacy of mathematics can be understood in several ways. It is not necessarily the science that should be taught first – grammar and logic should also be learned early in one's life, according to Roger Bacon – but it remains primitive, not only because of its antiquity but also because it is present in every kind of knowledge, in one form or another.

It is noteworthy to observe that Roger Bacon does not venture as far as advocating for a pure mathematical Platonism. It is true, he agrees, that human beings can naturally engage in mathematics even before being acquainted with its technical vocabulary and method. Yet, according to Bacon, what is initially inherent in us is logical reasoning. Not that logic as a science is innate, but rather, we naturally engage in rational procedures, subsequently formalised by the science referred to as logic.[26] Quoting Aristotle, he asserts that even uneducated individuals, even those deemed as simpletons, sometimes reason quite effectively.[27] Furthermore, quoting Alhazen this time, he says that practised logicians might not know where to seek the resolution of a problem within the *Organon* and yet manage to solve it.[28] A part of what we call "logic" would therefore be natural, even though it can be formalised (and therefore artificialised) as a science.

If logical reasoning is natural and innate, it is perfected first by reasoning, not by learning the science of logic, according to Roger Bacon.[29] It is therefore in scientific practices themselves that it can develop and become a *habitus*. Consequently, what is true for logic in general is also true for mathematical reasoning in particular, as Roger Bacon suggests in his commentary on Aristotle's *Metaphysics*. Here he imagines the possibility of teaching arithmetic without any other science than arithmetic itself:

> It is asked, if there is an essential and universal method, whether it must be acquired by a science. It seems that this is not the case, since the arithmetician, by teaching his disciple what is proper to his science through his own science, can explain its nature to him, by using a name, dividing, defining and collecting various elements, and thus inform him and teach him the demonstrative method of arithmetic without any other science.[30]

Roger Bacon does not agree with this argument. He does not deny that a science can be taught in this way, without systematically spelling out a discourse of method, but for us to be able to talk about science, we need to know the rules that allow our reasoning to be valid.

It is precisely this conception of a dual logic – one natural or innate and the other institutionalised as a science – that enables him to address the paradox of the *Meno*.[31] In this text, Roger Bacon demonstrates that one must posit something akin to a method common to all the sciences (he calls it "substantial" or "universal" method), but also particular methods, which depend on the matter to which it is applied. However, the question then arises concerning the science of this general method: how does one learn this method without using this very method? The natural character of logic allows for the rationale behind the quasi-innate presence of such reasoning even before the existence of logical science. We already have some dispositions and skills to think rationally and apply our reasonings to different matters.

Yet, this is insufficient, and Roger Bacon distinguishes a general method for all the sciences, which is natural but which can be acquired through the science called "logic" (which gives the rules for a science to be a science), alongside specific methods for each science (given, for instance, that moral philosophy and mathematics do not demand the same rigour, they do not use the same parts of logic).[32] He further specifies that the overarching scientific method is not solely logic; it necessitates the addition of metaphysics when positing principles within things.[33]

Naturally, this holds true for mathematics, elucidating how mathematics can be quasi-innate while subsequently evolving into a distinct science with its own rules and vocabulary. Indeed, the mathematical method corresponds in an eminent way to what Aristotle calls "demonstration" in the *Posterior Analytics*. According to Roger Bacon, this means that the science of logic requires mathematics in order to conceptualise what is the most excellent sort of demonstration:

> However, an understanding of logic is dependent on mathematics not only because of its end but also because of its middle and heart, which is the *Posterior Analytics*, for that book teaches the art of demonstration. But neither the premises of a demonstration nor its conclusions, nor the whole demonstration itself can be known, except with the principles of mathematics, because there alone is demonstration true and efficacious, as all know and which shall be explained later. Therefore, logic necessarily depends on mathematics.[34]

Therefore, the science of logic would not be complete without mathematics. In the *Communia mathematica*, Roger Bacon adds the following remark:

> I therefore assert that the principles of demonstration and the method for proving cannot be apprehended from the *Posterior Analytics* unless mathematical demonstrations are incorporated. And nothing of merit can be understood unless the scholar is versed in both geometry and arithmetic. For definitions, properties, axioms, postulates, and all the most worthy components cannot be gleaned from the *Posterior Analytics*, for they cannot be found in this book [....] Thus, one unacquainted with mathematics is incapable of perusing and instructing in the *Posterior Analytics*.[35]

As for the relationship between mathematics and metaphysics, it resides in the realm of quantity, understood as an accident of a substance. In the same way, metaphysics would not be complete without a mathematical explanation of quantity.

Roger Bacon thus does not assert that the human mind inherently possesses a stockpile of mathematical knowledge, thereby accounting for its primacy. Instead, he argues that mathematical reasoning, notably those of arithmetic, are not only among the oldest forms of logical reasoning in the history of scientific practices but also the most elementary and pervasive, because of their method and object. For these reasons, mathematics is present in almost every science, even in logic and grammar.

4.3 The Applicability of Arithmetic

In the second section of the fourth part of the *Opus maius*, Roger Bacon elaborates on the primacy of mathematics through its applicability to other sciences concerning material realities, followed by those concerning immaterial and divine realities.[36] First of all, the absence of mathematics would leave us ignorant of the workings of the heavens and their relationship with our sublunary world. It is primarily through geometry and *perspectiva* that we can grasp these actions. This understanding of the heavens also enhances our comprehension of the diverse climates, both in an astronomical and geographical sense, that constitute our world. This helps us to understand cultural and religious diversity, according to Bacon.

On the theological front, after reiterating the praises of mathematics proclaimed by the Church Fathers – especially Augustine – Bacon demonstrates how this knowledge is indispensable for interpreting Sacred Scripture. This is particularly the case with astronomy, insofar as it helps interpret locations and dates, whereas geometry and *perspectiva* would assist us to elucidate some phenomena described therein, or guide our contemplation of mysteries such as the Trinity.

Within this exposition of the *Opus maius*, arithmetic assumes a role seemingly less significant than the other sciences of the quadrivium.[37] It allows us to count various matters, which, according to Roger Bacon, hold particular utility in theology (a facet he expounds upon more extensively than its application in our worldly matters). The science of numbers, he explains, can assist us in passing from the literal to the spiritual signification of numbers mentioned in the Bible. It helps us to authenticate certain narratives, to interpret specific calculations, to rectify errors in manuscripts, to illustrate the Trinity, and even to describe concrete measurement systems occasionally referenced (such as the size of the Earth, Sun, or Moon, heights of clouds, etc.). In sum, it is important for biblical exegesis, but not as much for rational theology.

If the *Opus maius* was not strictly a theoretical treatise, but rather a political and religious text with a specific agenda, it stands in contrast to the

Communia mathematica. The latter can be seen as an extension of the *Opera* sent to the Pope, presenting a more precise and technical nature.

4.4 The Division of Mathematics in the *Communia mathematica*

In the first part of the *Communia mathematica*, we find a number of elements already present in the *Opus maius*. For instance, when Roger Bacon describes the origin of human errors in mathematics, he affirms that few people are interested in mathematics because they do not see its *utilitas*. But he aims, in this instance, to delve into more comprehensive elaborations, seeking to meticulously delineate the utility inherent in the various branches of mathematics. To do so, he classifies mathematical sciences according to the same procedure used in the *Communia naturalium* for natural philosophy.

After he gave a rather classical description of the quadrivium, quoting Boethius and Cassiodorus at length, he distinguishes within mathematics a part that concerns what is common to all mathematical sciences, a kind of methodological discourse, which seems to correspond to logical reasoning in mathematics, and a second part that corresponds to the quadrivium itself, but which is, like all other sciences, subdivided into two sub-parts: speculative and practical.[38] Mathematics is therefore divided in nine different parts (see Figure 4.1)

Generally speaking, mathematics deals with quantity. Geometry focuses on lines, surfaces, and solids; arithmetic on numbers; astronomy on the quantity of the stars and their relationship to terrestrial things; music on the quantity found in the sound of the voice and instruments. What Roger Bacon calls here general science or common knowledge in mathematical sciences – following al-Fārābī – is precisely this knowledge of quantity and related topics based on general definitions and distinctions (for instance, simultaneity, contiguity, continuity, position, place, space, infinite, unit, point, line, surface, solid, angles, curve, concave, convex, etc.).[39] Here we encounter again the duality between universal and particular methods within the realm of sciences, albeit at a further reduced level. Just as there exists a universal method applicable to all sciences, alongside specific methods unique to each, similarly, within each individual science, there exists a general component in mathematics along with specialised components. Roger Bacon compares this general or common part to what Euclid wanted to do in his *Elements*, although he deems it necessary to add a number of new things, particularly from the treatise *De sphaera* and the latest developments in mathematics.[40]

If one must acquire the common knowledge of mathematics, which is furnished by logic and metaphysics, as we know now, in order to engage with any of the four sciences of the quadrivium, there is, furthermore, a prescribed sequence to be followed in the apprenticeship of these disciplines.[41] Both astronomy and music necessitate familiarity with arithmetic and geometry. Given that music is the most intricate discipline, it is to be studied last. Additionally, one should begin with geometry before delving into arithmetic, adhering

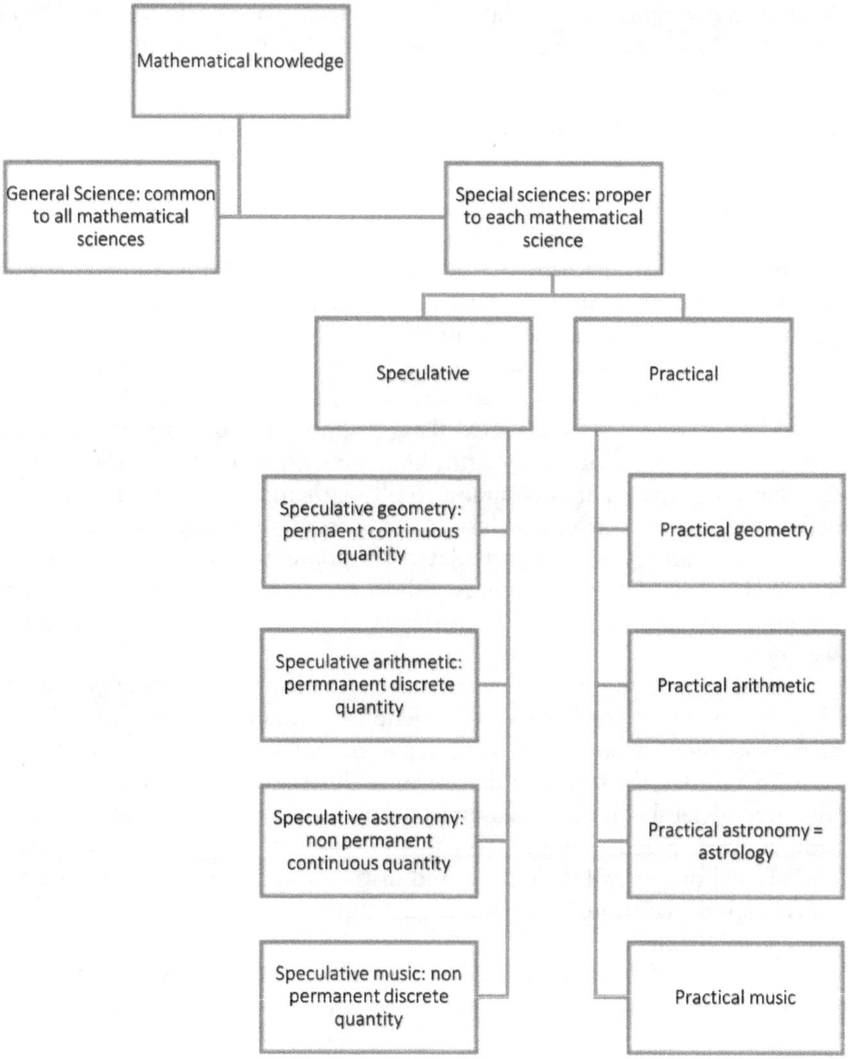

Figure 4.1 Flowchart Showing the Division of Mathematical Knowledge into Different Types of Sciences, Allowing One to Differentiate between Speculative and Practical Sciences.

to the guidance provided by metaphysical teachings, as well as the method employed by Euclid in the *Elements*. Indeed, contemplation through figures is more straightforward than through numbers; it resonates more closely, asserts Bacon, with our sensible knowledge. We shall later revisit the metaphysical presuppositions underlying this arrangement. First, let us explore how geometry and arithmetic are subdivided into theoretical and practical components.

Regarding practical geometry, it constitutes both the study of geometrical instruments (for measurement in particular) and the understanding of their applications, notably in civil life.[42] From this perspective, practical geometry pertains to the realm of agriculture (which, as we have observed, is also a distinct science within natural philosophy), owing to the knowledge encompassing isoperimetry, altimetry, planimetry, and bathymetry. It extends to architecture, naval, and river engineering, as well as the art of crafting machines for transport, conveyance, load-bearing, or killing enemies. Each of these practical sciences is learnt through classical texts, according to Bacon, such as Archimedes' *De isoperimetria* or Vitruvius' *De architectura*.

4.5 The Division of Arithmetic

Speculative arithmetic is concerned with numbers and their properties. According to Roger Bacon, the principles of this science of numbers are to be found in Euclid and Boethius, not in contemporary mathematicians like Jordan of Nemore, supposedly the leading light of more recent mathematics, whose theory is in fact based on the Boethian arithmetic.

> Indeed, although Boethius does not seem to proceed by demonstrations like those used by Jordan, the power and science of Jordan's demonstrations derive from Boethius. This is why Boethius' *On arithmetic* is excellent and more useful than Jordan's book, because few numbers have to be demonstrated in the manner used by Jordan, as will become evident below, and this book therefore contains a great deal of curiosity and superfluous material.[43]

As demonstrated by the second section of the *Communia mathematica*, the aspect that Roger Bacon appears to regard as most pivotal concerns the theory of proportions. Even though there exist different kinds of proportions (arithmetic, geometric, and harmonic), they all require a foundational understanding of arithmetic. On the practical front, Roger Bacon primarily references the tools of the abacus and Pythagorean table, followed by algorithms, astrological and astronomical tables, the art of weights and mixtures, mercantile calculations, the art of measuring altitudes, longitudes, latitudes, and depths using numbers (complementary to practical geometry), as well as the game of rithmomachia (a medieval variant of chess).

This compilation of practical arithmetic sciences aligns, to a greater or lesser extent, with the potential applications of arithmetic in philosophy and theology as expounded in the *Opus maius*. For, as we have seen, for Bacon, arithmetic serves primarily to verify calculations and measurements in sacred texts. While Roger Bacon continues to invoke Boethius, speculative arithmetic is no longer truly perceived as the essential foundation for all metaphysical and theological inquiries. Within his system, arithmetic has transformed into a primarily utilitarian discipline. It can assist us in interpreting secular or sacred texts, yet its principal utility resides in its practical applications.

4.6 The Philosophical Significance of the Primacy of Geometry over Arithmetic

True to his word, Roger Bacon describes continuous quantities before discrete quantities, geometric figures before numbers. This, he points out, is the custom of mathematicians.[44] But he hastens to add that here he must break with Euclid, who begins by defining the point as that which has no part, like an atom. According to Bacon, the notion of a point cannot be understood without first understanding what a straight line was.[45] This is because he himself understands the notion of point in the manner of Aristotle, i.e., as the terminus of a line segment, not as an entity in its own right, even if it is only considered abstractly and mathematically. After defining lines, surfaces, and solids, he makes this very clear:

> Since the continuum is infinitely divisible, that is, into entities that are always divisible, the straight line is not composed of points, but of straight lines, any more than surfaces are composed of indivisibles, or bodies of atoms that are called indivisible bodies (or individuals), or surfaces of straight lines and bodies of surfaces and straight lines. Every straight line is composed of straight lines, every surface is composed of surfaces, and every body is composed of bodies.[46]

Roger Bacon was fiercely opposed to atomism in general and to mathematical indivisibilism in particular. And, as he clearly states in the *Opus maius* about the position of Democritus and Leucippus, "this error is totally eliminated by the power of geometry."[47]

Roger Bacon was one of the first in the West to use the arguments based on incommensurability in geometry formulated by Avicenna and transmitted in Latin by al-Ghazali.[48] According to him:

> the only stronger argument against this position is to show that the diagonal of a square is commensurable with its side, i.e. that they have a common measure, i.e. an aliquot part as a common measure, which is contrary to what Aristotle always taught.[49]

Indeed, this kind of geometrical criticism of indivisibilism can be found throughout his work, from the *Opus maius* to the *Opus tertium*, via the *Communia naturalium* and the *Communia mathematica*.[50] Beyond the intrinsic force of these arguments, the emphasis on the question of measurement shows that Roger Bacon rejects the idea that atoms – whether points or corpuscles – can play the role of absolute units of measurement in the same way as units in numbers:

> But the point must be known before the finite line, since it is in its definition. But the point is not to be described as that which has no part, although Euclid says so, since this is also true of the unit and many other things, but as the extremity of a finite line.[51]

A point is not an indivisible unit of space, but the end of a line. Consequently, when we associate numbers with continuous quantities, we do this by commodity to make them commensurable or, on the contrary, incommensurable with each other, but this number corresponds to a sum of units that are conventional (*ad placitum*), i.e., according to our will and our capacities, not to absolute and independent units. They are nothing but units of measurement, not units that correspond to the ultimate indivisible parts of things, as the "indivisibilists" affirm:

> Since continuous quantities are infinitely divisible, so that we do not find in their nature anything indivisible by which several quantities could have something in common (*communicare*), the continuous quantity considered in itself is not subject to this having something in common (*communicatio*) or not. Since, however, we must find quantities which have something in common (*communicantes*) and are commensurable (*commensurabiles*), and their opposites, in order to be certain of their measurement in things and for many other reasons, we must therefore equate continuous quantities with numbers. And since these numbers cannot be determined by nature, we shall give them voluntarily and by institution and according to our position (…) and thus, although there is absolutely no indivisible in continuous quantities, there may be an indivisible according to the senses and also from our point of view, although this is sensible.[52]

Measurement systems are therefore always conventional and in a sense arbitrary, he notes, because they vary from place to place. The Latins chose the barley grain as the minimum for measuring lengths, which multiplied by four times gives a finger, which multiplied four times gives a palm, and so on.[53] In other countries, notes Bacon, the system of measures can be totally different.

It is therefore easy to understand why Roger Bacon refuses to give too much importance to arithmetic, contrary to a long tradition of Platonic and neo-Pythagorean inspiration handed down in particular by Boethius and revived in the twelfth century and still present in Robert Grosseteste.[54] For Bacon, numbers are in no way the principles of things but are mere abstractions that can only be conceived from the division of the continuum. In no case, therefore, is it possible to establish a direct and symmetrical relationship between numbers (discrete quantities) and geometric or physical quantities (continuous quantities), as Robert Grosseteste endeavoured to show in his *De luce,* for instance.[55] While arithmetic is necessary to the other mathematical sciences and even to other forms of knowledge, as we have seen,

Bacon no longer sees it as the gateway to all created and uncreated truths, as Boethius thought. There is, therefore, a great deal of consistency between his presentation of mathematical education and his philosophy of mathematics. In this respect, he is opposed not only to his hero Robert Grosseteste but also to his near-contemporary John Pecham.

4.7 A Brief Comparison with John Pecham's View on Arithmetic

While it is clear that John Pecham's treatises on optics owe much to the work of Roger Bacon, he also took some ideas from Robert Grosseteste.[56] Even though he accepted the Boethian dictum that mathematics is a sine qua non condition for reaching all truth, created and uncreated,[57] he gave *perspectiva* a prominent place in the explanation of the functioning of the universe, while subordinating it to geometry and physics. The real difference with Roger Bacon lies in the fact that he also subordinates geometry to arithmetic in the way Boethius did and seems to defend a form of mathematical atomism that brings him closer to Robert Grosseteste.

For Pecham, as for Bacon, the science of *perspectiva* studies the diffusion of light rays through geometry, with the aim of exhibiting the laws that govern our apprehension of the world through the sense of sight. John Pecham adds that the *perspectiva* studies light rays as both mathematical and natural entities, whereas geometry sticks to the mathematical line.[58] In other words, geometry is a more abstract science than the *perspectiva* and, therefore, allows us to come closer to the truth of Creation. Indeed, like Boethius, he systematically places his theses on optics in a wider theological context, drawing on numerous scriptural and patristic quotations. From the very first lines, he quotes a passage from Augustine's *De libero arbitrio*, in which it is said that scholars must turn away from earthly vileness and ascend to "number and wisdom."[59] It is through creatures that man must reach God, "so that through created number we may ascend to uncreated [number] and through created light to the light of uncreated wisdom."[60] If human beings can reach God's plan through optics, because it makes geometric laws visible to us, they also need the science of geometry and even more arithmetic, insofar as the world has been initially created with numbers.

How does this work? In a *quodlibet* devoted to the diffusion of light, John Pecham addresses this question when discussing the Aristotelian thesis of the instantaneousness of the diffusion of light.[61] He argues against Aristotle that light is constantly produced (it is *in fieri*). But a *respondens* objects that this thesis would imply a form of indivisibilism (*et ita divisible ex indivisibilibus*), which seems impossible for the objector. In his reply, which is not absolutely clear, John Pecham seems to accept a part of this atomism, but tries to show that it does not imply that the diffusion of light is discrete or that motion is composed of indivisibles. To show this, he uses an example that will become standard in fourteenth-century debates on the continuum, that of a sphere rolling on a plane. The contact between the sphere and the tangent surface is an indivisible point, but the motion of the sphere on this surface is continuous, and, therefore, traces a continuous line, in which one can distinguish the existence of all the points of contact with the sphere. So motion itself is not discrete or successive, nor is space. Applied to the case of light, it can be said that it is continuously generated while being composed of indivisible points of light, as in Robert Grosseteste's *De luce*. The underlying idea, it seems to me, is, as with Robert Grosseteste, that the equivalence between points and units makes it possible to understand where numbers nest in geometric figures and, by extension, in physical bodies. In his *De numeris misticis*, John Pecham goes even further, saying that numbers exist in some way *in* real things.

As Barnabas Hughes has shown in his introduction, the analysis of numbers in *De numeris misticis* is based on a long textual tradition that stretches from Augustine to Bonaventure, via Boethius and Robert Grosseteste.[62] The use of these sources mixes mathematical analysis with symbolic interpretations of numbers in biblical exegesis, so that the treatise is divided into two distinct parts: one dealing with numbers in general, in the manner of a treatise on arithmetic, and the other applying this theory to a numerological analysis of Scripture. As in the *Tractatus de perspectiva*, this theological dimension appears from the very first lines of the text:

> The Lord Jesus arranged all things in number, measure and weight, as Wisdom says [XI, 21]. According to Augustine, he made all things spring from numbers, because he also created everything through unchangeable numbers. Moreover, he transmitted Your mysteries in the Holy Scriptures wrapped in numbers. I will therefore teach about simple numbers what you have taught me of the simple.[63]

In general, as John Pecham says, the *sapientes mundi* approach numbers from different points of view: as metaphysicians, as musicians, as geometers, and as arithmeticians.[64] Music deals with numbers in sounds, geometry with the measurement of measurable quantities, and arithmetic with all countable quantities. As for metaphysics, it tries to explore the essence of numbers in general. In the rest of the text, Pecham sticks essentially to arithmetic, even if he touches in passing on some of the important metaphysical questions related to the mode of being of numbers.

Concerning the mode of being of numbers, Pecham's position is very clear: number is not a mere fiction of the mind, a mere abstraction, a concept, and even less a convention to cut and measure reality:

> The number is divided into numbering number and numbered number. The numbering number is the mathematical number that exists in the things that can be counted, not in the soul that counts, as some people falsely imagine. Indeed, the mathematical number and the natural number, which is called numbered, do not differ, if not as the mathematical

> body and the natural body. Some have asserted that the point which has no part does not exist and that the line without width and the surface without depth do not exist in the nature of things, but only according to their mode of being in the imagination. But their error is obvious, for the body, if it is finite, does not have a body as its extremity, for that would go to infinity. So it has for its extremity a surface. Moreover, the surface does not have a surface as its extremity for the same reason. It therefore ends in a line without width. In the same way, a line, if it is not infinite, has for its extremity an indivisible point [....] Thus all mathematical things exist in natural things.[65]

The *numerus numerans* and the *numerus numeratus* perfectly match, not simply because one is abstracted from the other, but because these numbers describe the same reality from two different angles. Applied to bodies, the *numerus numerans* refers to the mathematical body, while the other, the *numerus numeratus*, refers to the natural body, and these two bodies are in perfect correspondence. So, just as *perspectiva* studies geometric entities both as mathematical entities and as natural entities, arithmetic also describes the deep mathematical structure of geometric entities and the deep ontological structure of natural bodies. If the lines that make up surfaces are composed of indivisibles, i.e., points, which are like units in space, as he seems to accept in the *quodlibet* quoted earlier, then the number in a natural body corresponds to the number of units that ultimately constitute and structure a three-dimensional solid.

We thus find in the *De numeris mysticis* the fundamental intuition of neo-Pythagorean mathematical atomism.[66] Numbers are both in ideas and in things, not by a relation of participation, as in Plato, not in the mode of a simple abstraction, as in Aristotle, but as principles inscribed in the matter of everything. Solids are then really composed of surfaces, surfaces of lines, and lines of points, and since points are units in space, these spatial realities can also be expressed in numbers, as Boethius already wrote in *De arithmetica*. Because numbers are inscribed in geometrical figures and because these correspond point by point to things in nature, it can be said that, from a metaphysical point of view, numbers are the foundation of natural reality.

4.8 Conclusion

Despite their use of identical sources and their shared respect for Boethius and Robert Grosseteste, Roger Bacon and John Pecham defended different theorisations of mathematics. One of the main differences lies in the place they reserve for arithmetic and in the consequences this has for their ontology of mathematical and physical objects. Indeed, their opposition is not simply a matter of attitude towards the problem of the continuum (divisibilism vs indivisibilism) but reveals much stronger underlying trends in their respective philosophy of mathematics concerning the nature and role of numbers.

Roger Bacon confines himself to a purely utilitarian and mainly exegetical use of arithmetic, whereas for John Pecham, this is only part of the role reserved for arithmetic. The reason for this seems to be that for Pecham, numbers are the primary elements of his metaphysics of Creation (God created the world "in number"), whereas for Bacon, who is more Aristotelian from this point of view, numbers are always abstractions, derived from the potentially infinite division of the continuum (our physical world is primarily geometric). For Bacon, anything can be numbered, but the choice of units will always be arbitrary and dependent on our will, i.e., on human ends. Numbers, therefore, do not give us access to the objective structure of reality. For Pecham, on the other hand, numbers are grounded in reality because there are ultimate units in reality, as in any whole number. These units exist in geometric figures and natural bodies. In such a theory, numbers are abstract entities not because they depend on our capacities or will but because they give us access to fundamental elements of the universe.

Roger Bacon's appreciation of geometry can be discerned very clearly in the countless pages he devotes to the science of *perspectiva*, as this science is not limited to optics, but can be used to establish truths about many forms of natural causality. But this view of geometry as superior to arithmetic is also related to his attachment to a form of empiricism, in which many concepts are, and must be, derived from experience. This is also the case in mathematics. And even if we might think that Roger Bacon adheres to a form of Platonism when he considers mathematical reasoning to be innate, this is not the case. We only have a disposition for mathematical reasoning that is in us, which is revealed through the exercise of specific knowledge, in and through experience. From this point of view, numbers and their related concepts within arithmetic are derived from our experience, and above all from our experience of geometrical objects (since geometry governs our apprehension of the sensible world, especially through vision). One might think that, as with some Platonists, this does not contradict the primacy of numbers, even from a metaphysical and theological point of view. The fact that we have to abstract discrete quantities (numbers) from continuous quantities (magnitudes) could be the sign of the superiority of numbers. But this is not the case for Roger Bacon, because the continuum, i.e., the whole of reality as it appears to us, is infinitely divisible. Any attempt to arithmetise the world can only be an arbitrary way of cutting up the world, i.e., counting objects of which we can already think abstractly what they are. We cannot count the points in a line, but only smaller lines with a precise quantity.

Roger Bacon thus broke with a Platonic and neo-Pythagorean trend that was still widespread in his time, as can be seen in the writings of Robert Grosseteste and John Pecham. Like most philosophers after him, Roger Bacon defended a theory of mathematical abstraction inspired by Aristotle. But I hope that I have nevertheless shown the originality of his position, notably in his way of conceiving the presence of practical mathematics in all types of knowledge, as a kind of universal form of reasoning, and at the same time, in his way of relativising the forms that calculations and measurements can take in different cultures.

Notes

1 Bacon, *Opus tertium*, 70. On the identity of John of Tynemouth and the other characters mentioned by Bacon, see Knorr, "John of Tynemouth," 293–330.
2 Bacon, *Opus maius*, vol. I, 108.
3 It should be noted, however, that in his *Compendium studii philosophiae* and *Compendium studii theologiae*, Roger Bacon emphasises the sciences of language and barely mentions mathematics.
4 Lindberg, *Theories of Vision,* 107–116; Lindberg, "Roger Bacon on Light," 243–275.
5 Agrimi and Crisciani, "L'utilitas dell'astrologia," 489–494; Hackett, "Roger Bacon on Astronomy-Astrology," 175–198; Panti, "An Astrological Path to Wisdom," 363–400.
6 Loewen, "Roger Bacon and the Science of Music and Preaching," 129–165; Van Deusen, "Roger Bacon on Music," 223–241; and Kedar, "Sound is Not Made of Rays," 141–158.
7 This is the case, at least, for philosophical psychology. See for instance, Hackett, "Agent Intellect and Intelligible Species," 149–166.
8 On this issue, see Hughes, "Franciscans and Mathematics (I)," 98–128; B. Hughes, "Franciscans and Mathematics (II)," 3–66; and more recently Robert, "Les mathématiques franciscaines," 123–177.
9 See Power, *Roger Bacon*, 113–115. See also Egel, "Roger Bacon's New Founding of the Sciences," 125–140.
10 See Hackett, "Roger Bacon and the Moralization of Science," 371–392.
11 Bacon, *Opus maius*, vol. I, 62–64. On this subordination of sciences to theology, see Panti, "The Theological Use of Science at the Oxford Franciscan School," 181–210; and Panti, "The Theological Use of Science in Robert Grosseteste and Adam Marsh," 143–163.
12 Molland, "Roger Bacon's Knowledge of Mathematics," 151–174.
13 Lindberg, "Roger Bacon and the Origins of *perspectiva,*" 249–268. More recently, see Mantovani, "The First of All Natural Sciences," 186–214. For more on *scientia experimentalis*, see Jeremiah Hackett's chapter in this volume.
14 Bacon, *Opus tertium*, 74. Nec mirum si omnia a sciantur per mathematicam, et omnia per hanc, quia omnes scientiae sunt connexae (ut superius dixi), licet quaelibet simul cum hac habeat suam proprietatem. Unde quaelibet habet potestatem in aliam, nec potest una sciri sine alia, ut praedictum est. Et necesse est omnia sciri per hanc scientiam, quia omnes actiones rerum fiunt secundum specierum et virtutum multiplicationem ab agentibus huius mundi in materias patientes.
15 Hackett, "Roger Bacon on the Classification of the Sciences," 49–63.
16 For a general presentation of medieval classifications of sciences, see Weisheipl, "Classification of the Sciences," 54–90.
17 Bacon, *Communia naturalium*, vol. I, 1–14.
18 Bacon, *Communia naturalium*, 6.
19 Bacon, *Opus maius*, vol. I, 99–100; Bacon, *Communia mathematica*, 11–12.
20 Bacon, *Opus maius*, vol. I, 100–103; Bacon, *Communia mathematica*, 12–14. For this aspect in particular, which is much more developed in the *Communia mathematica*, see Seller, "Aspetti della *quantitas continua,*" 37–52.
21 Bacon, *Opus maius*, vol. I, 102; Bacon, *Communia mathematica*, 14–16.
22 Bacon, *Opus maius*, vol. I, 103; Bacon, *Opus tertium*, 216. We do not detail here all the reasons given by Roger Bacon in this passage.

23 Bacon, *Opus maius*, vol. I, 117 (English translation by Dennis, *Roger Bacon's Mathematical Thought*, 64).
24 Bacon, *Opus maius*, vol. I, 103.
25 Bacon, *Opus maius*, vol. I, 105–106.
26 Bacon, *Opus tertium*, 210–214.
27 Bacon, *Opus tertium*, 210.
28 Bacon, *Opus tertium*, 214.
29 On logic as natural in Roger Bacon, see Brumberg-Chaumont, "Raison et logique au Moyen Âge" (forthcoming). I would like to thank Julie Brumberg-Chaumont for sharing the preprint of this chapter with me.
30 Bacon, *Questiones supra libros prime philosophie Aristotelis*, 54. Queritur si est modus unus essentialis universalis utrum debeat haberi per scientiam. Quod non videtur: quia arismeticus in instruendo discipulum proprium in propria scientia potest exponere ei quid est quod dicitur per nomen, dividendi diffiniendi et colligendi, et informare ipsum et dicere ei modum demonstrandi aristmetice sine omni scientia.
31 Bacon, *Questiones supra libros prime philosophie Aristotelis*, 54–58.
32 On this aspect, see Robert, "L'idée de logique morale aux XIIIe et XIVe siècles," 27–46.
33 Bacon, *Questiones supra libros prime philosophie Aristotelis*, 57.
34 Bacon, *Opus maius*, vol. I, 102 (English translation, 68).
35 Bacon, *Communia mathematica*, 15 (my translation). Dico igitur quod principia demonstracionum et modus demonstrandi sciri non possunt in libro *Posteriorum* nisi adducantur demonstraciones mathematice nec potest aliquid dignum intelligi nisi doctor geometriam et arismeticam saltem sciat. Nam que sunt diffiniciones et passiones et peticiones et supposiciones, et simpliciter et alicui, ac maxime dignitates, sciri non possunt in libro *Posteriorum* eo quod ibi non ponuntur.
36 Bacon, *Opus maius*, vol. II, 109–285.
37 Bacon, *Opus maius*, vol. I, 219–236.
38 Bacon, *Communia mathematica*, 39.
39 Bacon, *Communia mathematica*, 19–37.
40 Bacon, *Communia mathematica*, 38.
41 Bacon, *Communia mathematica*, 40.
42 Bacon, *Communia mathematica*, 42–47
43 Bacon, *Communia mathematica*, 47. Boetius autem, licet secundum modum procedendi non videatur sic demonstracionibus uti sicut Jordanus, tamen virtus et sciencia demonstracionum Jordani ex fontibus Boetii oriuntur. Unde optimus liber est *Arismetica* Boetii et utilior quam liber Jordani, quia pauca de numeris demonstranda sunt sub modo demonstrandi quo utitur Jordanus, ut inferius manifestabitur evidenter, et ideo in isto libro magna superfluitas et curiositas continetur.
44 Bacon, *Communia mathematica*, 23.
45 Bacon, *Communia mathematica*, 24.
46 Bacon, *Communia mathematica*, 27. Quoniam autem continuum est divisibile in in infinitum hoc est in semper divisibilia, linea non componitur ex punctis sed ex lineis, nec superficies ex aliquibus indivisibilibus, nec corpus ex athomis que dicuntur corpora indivisibilia (aliter individua), nec est superficies ex lineis, nec corpus ex superficiebus nec lineis. Sed linea omnis componitur ex lineis, et superficies ex superficiebus, et corpus ex corporibus.
47 Bacon, *Opus maius*, vol. II, 151.
48 Thijssen, "Roger Bacon," 25–34.
49 Bacon, *Opus maius*, vol. II, 151.
50 For the *Communia naturalium*, see Panti, "Natural Continuity and the Mathematical Proofs against Indivisibilism," 159–190.

51 Bacon, *Communia mathematica,* 24. Sed ante lineam finitam oportet punctum cognosci, qui in eius diffinicione cadit. Punctus vero non debet describi per hoc quod est cuius pars non est, licet hoc dicat Euclides, quoniam hoc convenit unitati et multis aliis, sed est terminus linee finite.
52 Bacon, *Communia mathematica*, 106. Quoniam autem quantitates continue divisibiles sunt in infinitum, ut non inveniatur ex eorum racione aliquod indivisibile quo possent plures quantitates communicare, ideo quantitas continua secundum se considerata non subicietur communicacioni et eius opposito. Quoniam tamen necesse est nobis invenire quantitates continuas communicantes et commensurabiles et earum oppositas propter mensurarum certitudinem in rebus et propter multa, ideo oportet nos quantitates continuas assimilare numeris; et quod non determinatur eis ex natura dabimus eis ex voluntate et institucione et posicione nostra. Nec repugnamus in hoc nature quia natura non contradicit huic posicioni, sicut nec determinat set permittit eam fieri a nobis cum necesse est, et ideo licet in continuis quantitatibus non sit aliquid indivisibile simpliciter, tamen potest esse indivisibile secundum sensum et eciam secundum posicionem meam, licet sit sensibilem.
53 Bacon, *Communia mathematica*, 77–79 and 107.
54 For general presentations of this tradition in the Middle Ages, see Albertson, *Mathematical Theologies*; Caiazzo, Macris, and Robert (ed.), *Companion to the Reception of Pythagoras.*
55 On numbers and measurements in Robert Grosseteste, see Lewis, "Robert Grosseteste and the Continuum," 159–187. It is worth noticing that this strategy became common for many other indivisibilists throughout the Middle Ages.
56 On this topic, see Lindberg, *John Pecham and the Science of Optics*, 10–11 and 24–29.
57 See the texts mentioned in Kibre, "The Boethian *De institutione arithmetica* and the *Quadrivium*," 67–80 (in particular 75–77).
58 Pecham, *Tractatus de perspectiva*, 24.
59 Pecham, *Tractatus de perspectiva*, 23.
60 Pecham, *Tractatus de perspectiva*, 23. Ut per numerum creatum ascendamus ad increatum et per lucem creatam ad lucem sapientiae increatam.
61 Pecham, *Quodlibeta quattuor*, I, q. 7, 20–23.
62 Pecham, *De numeris misticis*, 2–28, 333–383, and introduction.
63 Pecham, *De numeris misticis*, 333. Omnia Domine Iesu teste Sapientia, numero et mensura et pondere disposuisti. Omnia secundum Augustinum numeris respersisti, quia etiam per numeros incommutabiles omnia conditsti. Tua insuper misteria in Scripturis Sanctis sub numerorum velaminibus tradidisti. Ideo de numeris tradam simplicibus que me simplicem docuisti.
64 Pecham, *De numeris misticis*, 333–334. It should be noted that astronomy has disappeared from this new quadrivium in favour of metaphysics and that the *perspectiva* is not mentioned.
65 Pecham, *De numeris misticis*, 334. Dividitur quidem numerus per numerantem et numeratum. Numerus numerans est numerus mathematicus qui est in rebus numerabilibus, non ut quidem falso imaginantur in anima numerante. Non enim differunt numerus mathematicus et numerus naturalis qui dicitur numeratus, nisi sicut corpus mathematicum et corpus naturale. Dixerunt aliqui etenim punctum cuius pars non est et lineam cuius non est latitudo et superficiem cuius non est profundum in rerum natura non esse, sed tantum in ymaginatione secundum quod huiusmodi. Sed horum error per hoc patet quia et corpus si est finitum terminatur non ad corpus quia sic iretur in infinitum. Ergo ad superficiem: item, superficies terminatur non ad superficiem eadem ratione. Ergo ad lineam carentem omni latitudine: similiter et linea si non est infinita, terminatur ad punctum penitus indivisi-

bile. Ergo in omni corpore naturali habente superficies diversas, omnia ista sunt in acto suo existentia secundum essentiam et diffinitionem mathematicam, habente inquam superficies diversas. Quoniam spera habet tantum unam superficiem et ideo nec punctum nec lineam in actu. Sic igitur omnes res mathematice sunt in ipsis rebus naturalibus.

66 For the medieval versions of this atomism, see Robert, "Atomisme pythagoricien et espace géométrique au Moyen Âge," 181–206.

Bibliography

Primary Sources

John Pecham. *De numeris misticis*. Edited by Barnabas Hughes, *Franciscanum historicum*, 78/1–2 (1985): 2–28, 333–383.

John Pecham. *Quodlibeta quattuor*. Edited by Ferdinand Delorme and Girard J. Etzkorn. Grottaferrata: Collegium Saint Bonaventure, 1989.

John Pecham. *Tractatus de perspectiva*. Edited by David C. Lindberg. St. Bonaventure: The Franciscan Institute, 1972.

Roger Bacon. *Communia mathematica*. Edited by Robert Steele. *Opera hactenus inedita Rogeri Baconi*, fasc 16. Oxford: Clarendon, 1905–1940.

Roger Bacon. *Communia naturalium*. Edited by Robert Steele. *Opera hactenus inedita Rogeri Baconi*, fascs 2–4. Oxford: Clarendon, 1905–1940.

Roger Bacon. *Opus majus*. Edited by John Henry Bridges. *3 vols*. Oxford and Edinburgh: Clarendon, 1897–1900.

Roger Bacon. *Opus majus IV*. In Patrick Willard Dennis. Trans. *Roger Bacon's mathematical thought: a translation of part IV of the Opus majus with Introduction and Commentary*. Dallas: Unpublished Ph.D dissertation, 2011.

Roger Bacon. *Opus tertium* Edited and Translated by Nikolaus Egel. Hamburg: Felix Meiner Verlag, 2020.

Roger Bacon. *Questiones supra libros prime philosophie Aristotelis*. Edited by Robert Steele. *Opera hactenus inedita Rogeri Baconi*, fasc. 10. Oxford: Clarendon, 1930.

Secondary Sources

Agrimi, J. and Crisciani, Chiara. "L'*utilitas* dell'astrologia in Albumazar e Ruggero Bacone." In *Actas del V Congreso Internacional de Filosofía Medieval*, 489–494. Madrid: Editoria Nacional, 1979.

Albertson, David. *Mathematical Theologies: Nicholas of Cusa and the Legacy of Thierry of Chartres*. Oxford: Oxford University Press, 2014.

Brumberg-Chaumont, Julie. "Raison et logique au Moyen Âge." In *La Raison au Moyen Âge*, edited by Dominique Poirel. Paris: Vrin, 2023.

Caiazzo, Irene, Macris Constantinos, and Robert Aurelien, eds. *Companion to the Reception of Pythagoras and Pythagoreanism in the Middle Ages and the Renaissance*. Leiden and Boston: Brill, 2022.

Egel, Nikolaus. "Roger Bacon's New Founding of the Sciences in the Thirteenth Century." In *The Philosophy and Science of Roger Bacon. Studies in Honour of Jeremiah Hackett*, edited by Nicola Polloni and Yael Kedar, 125–140. London/New York: Routledge, 2021.

Hackett, Jeremiah. "Agent Intellect and Intelligible Species in Roger Bacon and John Pecham." In *Die Seele im Mittelalter: Von der Substanz zum functionalen System*, edited by Günther Mensching and Alia Mensching-Estakhr, 149–166. Würzburg: Koenighausen and Neuman, 2018.

Hackett, Jeremiah. "Roger Bacon on Astronomy-Astrology: The Sources of the *Scientia experimentalis*." In *Roger Bacon and the Sciences – Commemorative Essays*, edited by Jeremiah Hackett, 175–198. Leiden: Brill, 1997.

Hackett, Jeremiah. "Roger Bacon on the Classification of the Sciences." In *Roger Bacon and the Sciences – Commemorative Essays*, edited by Jeremiah Hackett, 49–66. Leiden: Brill, 1997.

Hackett, Jeremiah. "Roger Bacon and the Moralization of Science: From *Perspectiva* through *Scientia Experimentalis* to *Moralis Philosophia*." In *I Francescani e le scienze*, 371–392. Spoleto: CISAM, 2012.

Hughes, Barnabas. "Franciscans and Mathematics (I)." *Archivum franciscanum historicum* 76 (1982): 98–128.

Hughes, Barnabas. "Franciscans and Mathematics (II)." *Archivum franciscanum historicum* 77 (1984): 3–66.

Kedar, Yael. "Sound is Not Made of Rays: Roger Bacon's Rejection of Heavenly Music." In *Roger Bacon and Medieval Science and Philosophy: Studies in Honour of Jeremiah Hackett*, edited by Nicola Polloni and Yael Kedar, 141–158. London: Routledge, 2021.

Kibre, Pearl. "The Boethian *De institutione arithmetica* and the *Quadrivium* in the thirteenth century university milieu at Paris." In *Boethius and the Liberal Arts. A Collection of Essays*, edited by Michael Massi, 67–80. Peter Lang: Berne-Frankfurt-Las Vegas, 1981.

Knorr, Wilbur R. "John of Tynemouth alias John of London: Emerging Portrait of a Singular Medieval Mathematician." *The British Journal for the History of Science* 23/3 (1990): 293–330.

Lewis, Neil. "Robert Grosseteste and the Continuum." In *Albertus Magnus and the Beginnings of the Medieval Reception of Aristotle in the Latin West*, edited by Ludger Honnefelder, 159–187. Münster: Aschendorf Verlag, 2005.

Lindberg, David C. "Roger Bacon on Light, Vision, and the Universal Emanation of Force." In *Roger Bacon and the Sciences – Commemorative Essays*, edited by Jeremiah Hackett, 243–275. Leiden: Brill, 1997.

Lindberg, David C. "Roger Bacon and the Origins of *Perspectiva* in the West." In *Mathematics and its Applications to Science and Natural Philosophy in the Middle Ages*, edited by Edward Grant and John Emery Murdoch, 249–268. London: Cambridge University Press, 1987.

Lindberg, David C. *Theories of Vision from Alkindi to Kepler*. Chicago and London: The University of Chicago Press, 1976.

Lindberg, David C. *John Pecham and the Science of Optics. Perspectiva Communis. Edited with an Introduction, English Translation, and Critical Notes*. Madison, Milwaukee and Londres: University of Wisconsin Press, 1970.

Loewen, Peter V. "Roger Bacon and the Science of Music and Preaching." In *Music in Early Franciscan Thought*, edited by Peter V. Loewen, 129–165. Leiden and Boston: Brill, 2013.

Mantovani, Mattia. "The First of All Natural Sciences. Roger Bacon on *Perspectiva* and Human Knowledge." *Vivarium* 59/3 (2021): 186–214.

Molland, George. In *Roger Bacon and the Sciences – Commemorative Essays*, edited by Jeremiah Hackett, 151–174. Leiden: Brill, 1997. https://books.google.com/books?id=Gy3Vp7TurVUC&pg=PA157.

Panti, Cecilia. "An Astrological Path to Wisdom. Richard de Fournival, Roger Bacon and the Attribution of the Pseudo-Ovidian *De vetula*." In *Richard de Fournival et les sciences au XIIIe siècle*, edited by Joelle Ducos and Christopher Lucken, 363–400. Florence: SISMEL, 2018.

Panti, Cecilia. "Natural Continuity and the Mathematical Proofs against Indivisibilism in Roger Bacon's De celestibus (Communia Naturalium, II)."

In *Roger Bacon's Communia Naturalium. A 13th Century Philosopher's Workshop*, edited by Paola Bernardini and Anna Rodolfi, 159–190. Florence: SISMEL, 2014.

Panti, Cecilia. "The Theological Use of Science at the Oxford Franciscan School: Thomas Docking, Roger Bacon, and Robert Grosseteste's Works." In *The Franciscan Order in the Medieval English Province and Beyond*, edited by Michael Robson and Patrick N. R. Zutshi, 181–210. Amsterdam: Amsterdam University Press, 2018.

Panti, Cecilia. "The Theological Use of Science in Robert Grosseteste and Adam Marsh According to Roger Bacon. The Case Study of the Rainbow." In *Robert Grosseteste and the Pursuit of Religious and Scientific Learning in the Middle Ages*, edited by Jack Cunningham and Mark Hocknull, 143–163. Berlin: Springer, 2016.

Power, Amanada. *Roger Bacon and the Defense of Christendom*. Cambridge: Cambridge University Press, 2013.

Robert, Aurelien. "Atomisme pythagoricien et espace géométrique au Moyen Âge." In *Lieu, Espace, Mouvement: physique, métaphysique et cosmologie (XIIe-XVIe siècle)*, edited by Tiziana Suarez-Nani, Olivier Ribordy, and Antonio Petagine, 181–206. Turnhout: Brepols, 2017.

Robert, Aurelien. "Les mathématiques franciscaines et leur diffusion en Italie. Le cas du *De macrocosmo* de Marco Trevisano (Venise, XIVe siècle)." In *Les savoirs profanes dans les ordres mendiants (XIIIe-XVe siècle)*, edited by Joel Chandelier and Aurelien Robert, 123–177. Rome: École française de Rome, 2023.

Robert Aurelien. "L'idée de logique morale aux XIIIe et XIVe siècles." *Médiévales* 63 (2012): 27–46.

Seller, Fabio. "Aspetti della *quantitas continua* nei *Communia mathematica* di Ruggero Bacone", *Archivio di storia della cultura* XXXII (2019): 37–52.

Thijssen, Johannes. "Roger Bacon (1214–1292/1297): A Neglected Source in the Medieval Continuum Debate." *Archives internationales d'histoire des sciences* 34 (1984): 25–34.

Van Deusen, Nancy. "Roger Bacon on Music." In *Roger Bacon and the Sciences Commemorative Essays*, edited by Jeremiah Hackett, 223–241. Leiden: Brill, 1997.

Weisheipl, James A. "Classification of the Sciences in Medieval Thought" *Mediaeval Studies* 27 (1965): 54–90.

5 On Vision

Roger Bacon's *Opus maius* (V.1–2) from the Perspective of Alhazen's *Optics* (I–II)

Nader El-Bizri

5.1 Preamble

This chapter focuses on the *Perspectiva* part of the *Opus maius* (V.1–2) of Roger Bacon,[1] set in the context of interpreting it through the theory of direct vision of the *Kitāb al-manāẓir* (*Book of Optics* I-II)[2] by the Arab mathematician, astronomer, architect, and optician Alhazen (Ibn al-Haytham). The main emphasis will be placed on the theories of Bacon and Alhazen in terms of understanding the nature of visual perception, while also highlighting their mathematised physics of light, along with the physiology, neurology, and cognitive analytics of direct vision, and the associated conditions of the occurrence of inferential errors in sight.[3]

A full analysis of Part V of *perspectiva* in the *Opus maius* will require a lengthier study than what we aim at addressing in this context through the lens of interpreting the assimilation of key leitmotifs from Alhazen's theory of direct vision in Bacon's line of inquiry. It is vital in this regard to note that the focus on direct vision avoids going into the fuller technical details of the geometrised physics of catoptrics and dioptrics in the science of optics while investigating the physical, physiological, neurological, psychological, and cognitive aspects of visual perception. Our aim is to show the impact of the principal elements in Alhazen's theory of direct vision on Bacon's *perspectiva*, and how the latter offered diversions and elaborations of his own in this regard. This will allow us to highlight the nuances of the technicalities in Bacon's approach and demonstrate to what extent they resonate with Alhazen's theory of visual perception. This line of inquiry is ultimately based on the evidenced and documented propositions that figure in Bacon's *Opus maius* that have a direct appeal to Alhazen's *Optics*. Indeed, one of Bacon's main sources in the science of optics is Alhazen's oeuvre in its Latin translation from the Arabic *Kitāb al-manāẓir* (*Book of Optics*; rendered in Latin under the titles: *Perspectiva*, *De aspectibus*, or *Opticae*).[4]

5.2 Physiology and Neurology of Vision

Following Alhazen's line of inquiry in accounting for direct vision, Bacon investigates phenomena associated with the brain, the optic nerves, and the

DOI: 10.4324/9781003405702-6

anatomical structure of the eye. He discusses the principles of optics while emphasising the aesthetic experience of delight in the beauties that shine forth through vision, light, and colour, in addition to the epistemic and utilitarian significance of that science in the way that it studies these visual phenomena experimentally, while applying it to the investigation of all physical entities in the heavens (through astronomy) and on Earth (in physics). The impact of Alhazen is obvious in this regard, and it is articulated alongside Bacon's appeals to the oeuvres of Aristotle, Ptolemy, Avicenna, and Constantinus Africanus (*Opus maius*, Part V, Part I, First Distinction, Chapter I).

Bacon's *perspectiva* begins with thematic investigations that correspond with Volume I of Alhazen's *Optics*, focusing on the physiology of the eye and the neurology of the optical system. This is suggestive of a similar order in treating the theories of direct vision, as these figure much later in the seventeenth century in the synthesis of the science of optics within the mathematical oeuvre of the Jesuit Belgian architect and optician François d'Aguilon in his Antwerp school treatise, the *Opticorum libri sex*.[5] This opus contained enticing vignettes that were illustrated by the Flemish baroque painter, Sir Peter Rubens and engraved by the deacon Theodoor Galle; all done in illustration of the experiments that are described in Alhazen's *Optics*. These vignettes also depict the anatomical structure of the eye and the associated activity of dissection by figurines that represent the muses or angelic aids who serve the purposes of the experimenting optician, while also emphasising the frontal part of the brain (*al-dimāgh*), which for Alhazen would be the seat of the "last sentient organ" (*al-ḥāss al-akhīr*). A similar focus on Alhazen's investigation of the anatomy of the eye, the optic nerves (*ʿaṣab al-ʿayn*), the common nerve (*al-ʿaṣaba al-mushtaraka*), and the anterior part of the brain (*muqaddam al-dimāgh*) becomes the starting thematic features in Bacon's *perspectiva* studies. Alhazen deals with the structure of the eye in Chapter 5 of Volume I of his *Optics*. He engages in this after demonstrating his wider epistemic project of mathematising physics and the positing of the rudiments of the experimental and controlled-testing methodology. This underpins his demonstration of the separation of light from sight in the elaboration of his own theory of direct vision, and how it transcends the accounts concerning visual perception that preceded him (whether these are enshrined in the theses of Plato, Aristotle, Euclid, Ptolemy, the Stoics, and Galen in the Graeco-Roman traditions, or within the doctrines of figures like al-Kindī or al-Fārābī in the Arabic scientific lineage).

Alhazen aims at showing that direct vision does not simply occur through one series of phenomena, such as the intromission into the eyes of the rays of light that emanate from the outer lit surfaces of the object of vision that face the observer and pass in rectilinear trajectories in the isotropic and homogeneously transparent medium, such as air, which is between the object of vision and the observer. Nor does he only focus on the passage of these rays of physical light within the eye as they cross its anterior transparent membranes and refract as they cross the crystalline humour, which is itself not simply a lens but is essentially a sensing organ as well. The physiology of the anatomical structure of the eye in studying vision also leads to an examination of the optic nerves, and the common nerve, in a neurological account of visual perception, which reaches, in its subtle sensations, the last sentient organ (*al-ḥāss al-akhīr*) in the frontal part of the brain. All of these physical, physiological, neurological, and cerebral phenomena constitute the biological underpinnings of Alhazen's cognitive psychology of vision and his philosophy of mind.

From the onset, Bacon is keen on focusing his *perspectiva* on the common sense (*sensus communis*; αἴσθησις κοινὴ [*aísthēsis koinè*]; *ḥiss mushtarak*); while Alhazen deals with this aspect only after a long series of inquiries in geometrical optics, and through the studies in physiology and neurology of vision. Bacon places the common sense in the frontal part of the brain, which is where Alhazen locates the last sentient organ (*al-ḥāss al-akhīr*).

According to Alhazen, the crystalline humour (*jalīdiyya*) is focal in all this analysis as a locus of the transition from considering the eye as a nested series of lenses in catoptrics and dioptrics to looking at it as a living organ from the physiological viewpoint and the neurology of the optic nervous system to which it belongs. From a geometrised abstractive viewpoint, the crystalline humour is a dioptre instrument as a lens (*'adasa*). At the same time, it is physiologically a living, sensing, biological organ that feels the impress of the physical rays of light (*ḍaw'*) in the form of a subtle pain. Bacon follows this line of analysis and accentuates what Alhazen furthermore noted in this regard concerning the refined small filaments in the optical nerves that are filled with the same humour as the vitreous (*zujājiyya*), and which are internally located at the posterior organ in the eye after the crystalline humour.

There is a point-by-point correspondence – some sort of pointillism – between the points that are on the lit and visible surfaces of the object of vision, and the points of the sensory impress of the rays of light that are emitted from them and that enter the eyes of the observer. The image of the surface of the object of vision, which faces the observer, is correlated with the configuration of the point-by-point subtle sensations of the impress of the rays of light coming from that seen object. These eventually touch the nervous sensory optical system after they enter the eyes. This process starts with a sensation that is felt in the crystalline lens and is then transmitted in both eyes as a unified pattern of sensations that pass through the optic nerves and are ultimately felt by the last sentient organ in the front of the brain. This is because from any given point on the lit and visible surface of the object of vision, innumerable rays of physical light are irradiated spherically in all directions. However, only one single ray of light that is irradiated from that point falls perpendicularly as an orthogonal rectilinear trajectory on the outer transparent spherical membranes of the eye, and hence, passes through them without reflection or refraction to leave a single impress on the crystalline humour as a subtle sensation. This secures the point-by-point correspondence of the actual configuration of the lit visible surface of the object of vision and the pattern of subtle sensations it leaves as an impress of its light-rays on the optical nervous system of the eye. Seeing is, therefore, not simply a visual optical phenomenon, but it is haptic in the sense of having the light-rays from the object of vision touching the interior of the eyes in a physical manner, creating a sensation. The crystalline humour, as situated within the eye, gives us an experimental space that is akin to the installation of a *camera obscura* that has an anaclastic lens within it upon which light-rays are physically projected from the standpoint of a geometrised physics of light and dioptrics. The added aspect to consider herein is that the crystalline humour is also a sensing organ within a living biological eye, from the standpoint of physiology and anatomy, and is connected as such to an optical nervous system that is ultimately fused into the brain.

Understanding the intromission of the physical rays of light into the eyes requires a knowledge of dioptrics, and particularly of the anaclastic properties of spherical sections, as opposed to conics or the sections of the cylinder. This is the case given the overall approximation of spherical sections of the anterior membrane coats of the eye and the actual shape of the crystalline lens. A basic study conducted by Alhazen uses a spherical transparent glass vessel that is filled with water. This simulates the conditions of the spherical sections of the outer membranes of the eye when placed in a *camera obscura* and subjected to a controlled array of light rays. It can also act as an experimental model for a single droplet of rain in explicating the phenomenon of the rainbow (*Qaws quzaḥ*) as conducted in the *Tanqīḥ al-manāẓir* (*Recension of the Optics [of Alhazen]*) by the Persian optician Kamāl al-Dīn al-Fārisī,[6] or the contemporaneous study on the same legacy of Alhazen by the German Dominican Theodoric of Freiberg in the *De iride et radialibus impressionibus* (*On the Rainbow and the Impressions Created by Irradiance*),[7] all while considering the conditions of spherical aberration and the fuzziness in the capacity to focus the rays of light. This is nonetheless vital in terms of research on colours in connection with physical light, not only in the meteorological study of the rainbow.

The geometrising of the notions of physics into a structure of isomorphism between mathematics and natural philosophy (*qua* physics) acts as a basis for the mathematised structural modelling of the experimental controlled-testing procedures and installations as designed and devised by Alhazen. These are entwined with his science of optics in studying light and direct vision, as well as catoptrics and dioptrics. Experimentation is, as such, integral to the demonstration and verification procedures and methods in his *Optics*, while with Bacon, the experimental scientific procedure (*De scientia experimentalis*) is set within Part VI of his *Opus maius* and not in Part V on *perspectiva*.

Moreover, under the influence of the *De anima* of Avicenna (Ibn Sīnā), and the mediated medical observations of the compilations of the *Liber pantegni* (παντεχνῆ)[8] by Constantinus Africanus, Bacon seems to suggest that judgement as a cognitive process begins with the common sense, which he places in the same location where Alhazen situates the last sentient organ (*al-ḥāss al-akhīr*) in the frontal part of the brain. Bacon thus stresses the importance of studying the structure of the eye in the works by the authors of perspective, as well as the natural philosophers and the medical writers. He affirms that the three authorities in this are Alhazen, Avicenna, and Constantinus Africanus. He adds that they all agreed that there are two parts in the anterior cavity of the brain, which are called ventricles, or concavities. These cells are not the organs of the common sense and of the imagination, since those faculties are arranged anteriorly and posteriorly, while the ventricles of the brain are placed on the right and on the left, and in the anterior part as well. From the pia mater, which covers both ventricles, comes a double nerve; one from the right ventricle and the other from the left. These are the optical nerves, which then meet and are again divided (*Opus maius*, Part V, Part I, Second Distinction, Chapter 1).

As for the eye, it has three coats or membranes, three humours, and a web like that of a spider. Its first coat is formed from the interior coat of the nerve, which comes from the pia mater, and spreads out from the end of the nerve where it enters the opening in the bone and is ramified like a concave net. It is therefore called *rete* or *retina* (*shabakiyya*) and is supplied with veins, arteries, and slender nerves. Its second part is thicker and spreads out spherically up to the frontal part of the eye with an opening in the middle of its own fore part, in order that the impressions of light and colour may be able to pass through the middle of the eye to the nerve coming from the brain. This opening is placed directly opposite the extremity of the nerve at which the retina expands. Therefore, Alhazen states that this whole coat has two openings, one in front and the other behind, which is the extremity of the concave nerve. This second part is called the *uvea*, since it is like a grape (*'inabiyya*), because it has an opening in its anterior part like that left in a grape when its stalk is removed. This coat has two parts: the first is composed of veins, nerves, and arteries. The second spreads as far as the front of the eye, and a segment of it is visible as a portion of a sphere above the extremity of the *uvea* like a transparent horn, hence its name of "cornea" (*qurniyya*; from the root Arabic term *qurn* as "horn"). Despite its transparency, it has four thin, rind-like strong coats that shield the eye from the injuries and blows coming from the air. The third coat of the eye is formed from that third skin of the nerve that comes from the membrane of the skull, and its first part is joined to the bone of the eye and is hard and solid, and it is called *consolidativa* or *conjunctiva* (*multaḥima*). Bacon then adds that Alhazen says that the *uvea* has two openings, one anterior and the other posterior, which is the opening of the nerve, at which the expansion of the concavity of the *uvea* begins. There is also a humour that is *glaciale*, *crystallinum*, or *grandinosum*, which is like ice, crystal, or hail, and is formed at the extremity of the optic nerve. It has two parts; one is interior at the extremity of the nerve and is like melted glass, and on this account is called the *vitreus* humour, while the other part is anterior and is like ice, hail, or crystal, and is whiter than the vitreous humour, and is called anterior *glacialis*. There is ultimately the cornea (*qurniyya*), the humour *albugineus* (*al-ruṭūba al-bayḍiyya*), the humour *glacialis* (*al-ruṭūba al-jalīdiyya*), the humour *vitreus* (*al-ruṭūba al-zujājiyya*), and the extremity of the optic nerve (*'aṣab al-'ayn*). Consequently, the impressions of the rays of light will pass through their media to reach the front of the brain (*Opus maius*, Part V, Part I, Second Distinction, Chapter 2).

Unlike the pathway that is undertaken by Alhazen in terms of his analytics of the geometrised physics of light, the examination of the properties in catroptics and dioptrics of the anatomical elements of the eye, which lead to the focus on the neurological, cerebral, and cognitive aspects of vision, Bacon begins with the common sense and its location in the anterior part of the brain and then proceeds to the psychology of visual perception. Bacon focuses on the connection of the brain to the common nerves and how these move outwards through the optic nerves that pass via the aperture in the bone cavity that houses the eye. He then examines the various elements of the anatomy of the eye, such as the vitreous humour and the crystalline lens, and the frontal transparent membranes. Bacon begins from the location of the last sentient organ in the front of the brain and proceeds in his examination outward to reach the outer membrane of the eye as an interface with the ambient visible environment, while Alhazen begins with the lit and visible surfaces of the object of vision and proceeds inward towards the intromission of the rays of light in the eyes and the passing of their physical impress as sensations through the optical nerves to reach the anterior part of the brain.

5.3 Psychology of Vision

Bacon focuses on the role of the common sense in the cognitive aspects of visual perception and its interplay with the faculties of the mind, emphasising the imagination of *phantasia* and the recollection in memory. He also evokes the role of the faculty of estimation or prehension (*al-wahm*; *aestimativa*)[9] as it was elaborated by the philosopher and physician Avicenna, which designates a faculty that perceives the insensible forms that relate to sensible matter. There are in this line of analysis two classes of senses. The first class of senses is external and comes through the sensory experiences from the ambient outer environment, which principally evokes the function of the common sense that is located in the anterior part of the brain (according to Bacon). The second class of senses is internal, and it pertains to aspects of judgement in visual perception, with a particular emphasis on the estimative faculty and the manner with which it detects in the object of vision a given intentional meaning that results in fear. This entails the workings of a cognitive process that derives a certain meaning from an external sensorial stimulus that signifies some form of danger – like the sheep fearing the wolf upon seeing it. The intentional meaning that is detected in that experiential context as a manifestation of a source of fear, as it is associated by the estimative faculty with a given external entity in the outer ambient environment, can itself be akin to other forms of experiences that are aesthetic in their qualities. The estimative faculty is found not only associated with human consciousness and cognition but is also an attribute of the inner sense of certain animals. These animals display an expression of an internal processing of external sensory stimuli akin to sentience and explain how some animals react to an impending source of external danger. They process the intentional meaning of danger as identified with the visible properties of the object of vision that poses the threat, facilitating their flight from that given context.

In discussing these inner workings of the faculties by way of evoking family resemblances with the behaviour of certain animals that borders on some expression of sentience, this analysis is tied to the common sense and its connections with the imagination and memory. Even though Bacon appeals to Avicenna in this context, the analysis of the faculties of cognition in terms of discernment, syllogistic measure, imagination, and memory is marked by a reliance on Alhazen's *Optics*, not only in terms of direct vision but also in studying the inferential errors of visual perception. Alhazen's consideration of the last sentient organ highlights the close connection of the cerebral organs with the associated psychological and cognitive manifestations of the functional workings of the mind and its various faculties. Alhazen did so without determining the locations of the activities of the faculties in specific parts of the brain; the notable exception being the last sentient organ that he locates in the front of the brain. This last sentient organ becomes the psychosomatic locus of the transition from the physiological and neurological consideration of the processes of visual perception to the psychological and cognitive expressions of the mental workings of the faculties.

The physiological and neurological processes of visual perception that begin with the felt sensations in the crystalline humour lens end in the last sentient organ. The transmission of the sensory physical impress of the rays of light on the crystalline humour is transferred into the vitreous humour and the filaments of the optic nerves, which then become conjoined as a unified sensation in the common optical nerve, and then split again into two nerve conduits that reunite in the last sentient organ in the anterior part of the brain. The final sensory impress that is felt by the last sentient organ is transformed further into what is attributable to psychological and cognitive processes of the interplay of the mental faculties, without locating where the activities of the workings of discernment, judgement, imagination, or memory are located within the brain. The mental functions have a logical character to them that is associated with syllogism (*qiyās*) in a form of demonstrative thinking that arrives at its judgement by identifying the visible properties through the intentional meaning that is mentally associated with them. This evokes Alhazen's philosophy of mind that is anchored on a neuro-cognitive science and its cerebral analysis, albeit without speculating about the location in the brain of these mental activities.

The interplay between the common sense, imagination, memory, and estimation, according to Bacon, is a fortiori further elaborated by Alhazen in accentuating the role of discerning judgements that have the logical form of syllogistic demonstration in appraising the intentionality of signification as associated with the meaning of the visible properties of the objects of vision. The physiological and neurological analytics underpin the psychological and cognitive aspects of vision. The analysis of the anatomy of the eyes and the optical nervous system, along with the examination of the brain, are all undertaken as a basis for understanding the physiological and neurological processes of vision, and by way of setting the ground for engaging in the psychological and cognitive aspects of visual perception and the presuppositions of a psychosomatic philosophy of mind in Alhazen's theory of visual perception.

Alhazen lays down the groundwork in the science of optics for elaborating his philosophy of mind through his analysis of visual perception. He starts with the conditions of the geometrised physics of light in its spherical irradiation from every point on the lit surfaces of a given object, and how this emits innumerable secondary accidental rays of physical light that propagate in rectilinear trajectories across the space of an isotropic homogeneous transparent medium (such as the body of the ambient air between the outer lit and visible surfaces of the object of vision and the eyes of the observer who faces them). He then studies the conditions of the intromission of the rays of secondary accidental lights that are coming from the outer lit and visible surfaces of the object of vision and how they enter the eyes of the observer who faces them. This analysis evokes the conditions of reflection and refraction, in such processes of intromission as the rays of light pass through the frontal transparent membranes of the eye that differ in their refractive indices. This process reaches within the eye to the crystalline lens through which the rays of light are refracted. The crystalline lens here also acts as a sense organ to carry subtle sensations of the impress of physical light to the vitreous humour and the filaments of the optical nerves, passing behind the eye-cavities in the skull to the common nerve, then finally through to reach the last sentient organ.

Bacon also accounts for the workings of the faculties of imagination and the common sense from the standpoint of the neurological function of the optic nerves and their connection to the frontal part of the brain. In anatomical descriptions, the visual nerves descend from the double membranes of the brain; namely, the soft and tender *pia mater*, which is akin to a marrow ointment that enfolds the brain by direct contact, with phlegm as its main constituent, and the harder membrane *dura mater*, which adheres to the concave side of the lining of the cranium and resists the impress of the bone, with more hardened phlegm as its chief constituent. The soft matter of the brain has three distinctions as chambers or cells. The anterior part of the first cell is the seat of the common sense (*sensus communis*). This is like a fountain with respect to the senses, since judgement is incomplete before the visible form of the object of vision is processed by the common sense that distinguishes between the impressions on sight in categories.

Another faculty of the soul is situated in the posterior part of the first cell, the function of which is to retain the forms coming from the senses, owing to its tempered moistness and dryness, and this is the imagination, and is the repository of the common sense. The composition of the functions of the imagination with those of the common sense constitutes *phantasia* or the *virtus phantastica*. The common sense receives the form, imagination retains it, and the full judgement is reached as such by *phantasia* (*Opus maius*, Part V, Part I, First Distinction, Chapter 2). Accordingly, the imagination is the storehouse of the common sense, memory retains the impress of estimation, while cogitation orders all these faculties. In this regard, Bacon appeals to Avicenna more than Alhazen. Bacon also states that the estimative faculty takes the place of reason in brutes in dealing with internal senses that are hidden from the external senses, as the quality of a harmful or useful complexion, or rather the essential nature itself, whether useful or harmful. The forms or species that are in the imagination multiply themselves by way of the workings of the cogitative faculty, although they exist in the imagination according to their nature due to *phantasia*, while the cogitative faculty of the rational soul uses all the other faculties as its instruments, wherein species are categorised in the rational soul (*Opus maius*, Part V, Part I, First Distinction, Chapter 4).

5.4 Visible Intentional Properties of the Objects of Vision

In accounting for the object of vision, Alhazen refers to 22 visible intentional properties as *maʿānī mubṣara* (*intentiones visibiles*; *Optics*, II.3). According to him, these are: physical primary and accidental light, colour, distance, position, solidity, shape, size, separation, continuity, number, motion, rest, roughness, smoothness, transparency, opacity, shadow, darkness, beauty, ugliness, similarity, and dissimilarity. Alhazen then investigates the ways in which various errors in visual perception occur due to the cognitive erring in inferences that are made by the observer when the principal visible properties of distance, position, illumination, size, opacity, transparency, duration of perception, and the natural conditions of the eyes each falls outside its moderate range and cause its specific form of error in sight (*Optics*, III.7). Similarly, Bacon notes that the imagination and the common sense form judgements of themselves only concerning the intentional visible properties; namely, distance, position, figure, magnitude, continuity, discreetness or separation, number, motion, rest, roughness, smoothness, transparency, thickness, shadow, darkness, beauty, ugliness, as well as the similarities and differences in all the things composed of them. Besides these attributes, there are some that are placed under one or more of these qualities, as order under position, or writing and painting under figure and order, etc. Further examples are given regarding straightness and crookedness, or concavity and convexity, which are placed under the property of the figure. As for multitude and fewness, they are placed under number, while equality, augmentation, and diminution are placed under similarity and difference. In addition, eagerness, laughter, sadness, and lamenting are apprehended from the signifiers of the changes in the shape of the face. Moreover, moistness and dryness are placed under motion and rest. According to Bacon, all these properties, along with those that are reduced to them, are apprehended by the common sense and imagination (*Opus maius*, Part V, Part I, First Distinction, Chapter 3).

The visible properties of the object of vision are detected by the observer and interpreted inferentially as carrying specific attributes and meanings that facilitate their recognition and identification, or at times that result in visual errors, or *trompe l'oeil* illusionary phenomena, when falling outside the optimal moderate range. According to Alhazen, this entails that vision is a cognitive process and is not simply reducible to a physiological or neurological response to the effects of the physical rays of light that enter through the eye and leave an impress on its optical nervous system when reaching the brain as physical sensations. There is rather a psychosomatic character to vision in such a way that visual perception cannot occur without a recognition of the intended meaning and even the naming of the visible attributes that are associated with the object of vision. Visual perception is entwined as such with a philosophy of mind and cognition, and it ultimately designates a logos-based capacity of naming the attributes of what is seen.

Alhazen's philosophy of mind, which is articulated through his analysis of the cognitive aspects of visual perception, reflects an emphasis on recognition and prior knowledge. It is not simply an innate natural capability in the inner senses and the workings of the faculties but is also rooted in nurture and rearing. It is therefore a cognitive aspect that relates to learning and the acquisition of knowledge in an intersubjective milieu. Cognition is, therefore, an actualisation of the ability to synthesise prior knowledge by bringing what is unknown to the sphere of becoming known by way of recognition. This relies on a theory of species that is arrived at with some form of syllogistic reasoning, which classifies what is barely known under a given kind that allows, through recognition, to name its attributes and intentional meanings. Nothing in the visual field can stand out as being utterly outside the sphere of being classed and named; hence, being classified and recognised, and ultimately being brought under the cognitive fold of understanding and logos. Anything that is unnameable and unrecognisable can become a source of perplexity or even of fear and anguish. Even if its particularity and singularity as a visible phenomenon are not recognisable, it can still be classified under a given species through which it becomes identifiable and is hence assigned an intentional signification and meaning.

5.5 Phenomenology of Vision

The psychological and cognitive aspects of visual perception are moreover tied to a phenomenology of embodiment and inter-corporeal spatiality in the context of Alhazen's analysis. This is the case given that the kinaesthetic ocular-motor aspects of vision require an embodied situation of the observer who measures the visual field through the scale of their own corpulence, and even in the pace of the feet in walking, wherein they are connected to the ground as a common terrain with the contiguous and continuous interconnection of the surfaces of the visible objects of vision.

Besides the focus on the cognitive psychology and philosophy of mind in the analysis of visual perception, and the intersubjective realm of its logical determination of the intentional meaning of the visible properties of the objects of vision in their situational visualisation field, the emphasis on embodiment is evident and clear in Alhazen's phenomenological descriptions. The observers are embodied, and they measure the visual field through their own corporeal attributes and kinaesthetic ocular-motor comportment in inter-corporeal and intersubjective relations. This places an emphasis on the situational immersion of the observer in the visual field with its spatial and temporal apportioning of inhabiting it and disclosing its features through displacement. This can be in the shift of the gaze of the eyes, the movement of the neck and of the whole body in crossing over the visual field, or in an interplay of the unveiling and veiling of its multiple facets, wherein what is unconcealed hides again what was visible prior to the bodily motion. For example, if I take an opaque cube and hold it in one of my hands, then, if I look frontally at only one of its square-shaped sides, its five other faces are concealed. If I look at it with a twist, only three of its six sides appear at the same time, and if I afterwards rotate it, these three sides become concealed, while the other previously veiled three sides become visible. Even with a small object of vision that has a pure formal property of being a cube, we have this level of the intricacy of the space-time continuum in the kinaesthetic ocular-motor engagement of the embodied moving observer with the interplay of veiling and unveiling, concealment and un-concealment, and visibility and invisibility, in the physical presence of things. This becomes far more complex with a wide visual field and its horizons and the multitude of objects of vision within it, and spans of distances in space and time that the embodied observer must cross in view of disclosing its manifold visible intentional properties.

5.6 *Perspectiva Artificialis*

Overall, Alhazen offered a *Gestalt* theory of vision in his analysis of the workings of the mind in discernment (*tamyīz*), syllogistic comparative measure (*qiyās*), and inferential judgement (*ḥukm*), which amount to the recognition (*ma'rifa*) of what is seen, while also evoking in this context the interplay between memory (*dhikr* or *ḥāfiẓa*) and imagination (*takhayyul* or *khayāl*). This relates to the visual perception, via inferential judgement and recognition, of 22 intentional properties (*ma'ānī mubṣara*) that cover the cognisance of the light and colours of the objects of vision, their distance, position, solidity, shape, size, and separation from other things or continuity with them, the numbering of what is seen, its motion or rest, roughness or smoothness, transparency or opacity, shadows and darkness, beauty or ugliness, similarities or dissimilarities. All these visible intentional properties show that direct vision occurs by way of cognition, not simply in terms of ocular sensations in physiological and neurological terms, and in articulating a philosophy of mind and cognitive psychology that is also grounded on a phenomenology of situated lived ocular-kinaesthetic experiences in embodiment. These phenomena were later applied in the architectural and visual *cum* plastic art and spaces of the *Quattrocento* Italian Renaissance, especially through the Florentine adaptation of Alhazen's *perspectiva naturalis* in terms of projecting the *perspectiva artificialis*. After all, Alhazen was an architectural engineer (*muhanddis*) besides being an optician, astronomer, and geometer, and his transmitted architectural sensibilities in optics resonated as such with their adaptive interpretation through the *Della Pittura* (*De Pictura*; *On Painting*) of Leon Battista Alberti and the "Commentario terzo" ("Third Commentary") in the *Commentarii* (*Commentaries*) of Lorenzo Ghiberti.[10] This turn in the history of ideas and material culture refracted the trajectories of science with those of the visual arts and architecture via novel synergies and confluences that were articulated in the development of *perspectiva* in theory and practice, and as these found further resonances amongst others in François d'Aguilon's *Opticorum* and Girard Desargues' *Manière universelle pour pratiquer la perspective*.[11]

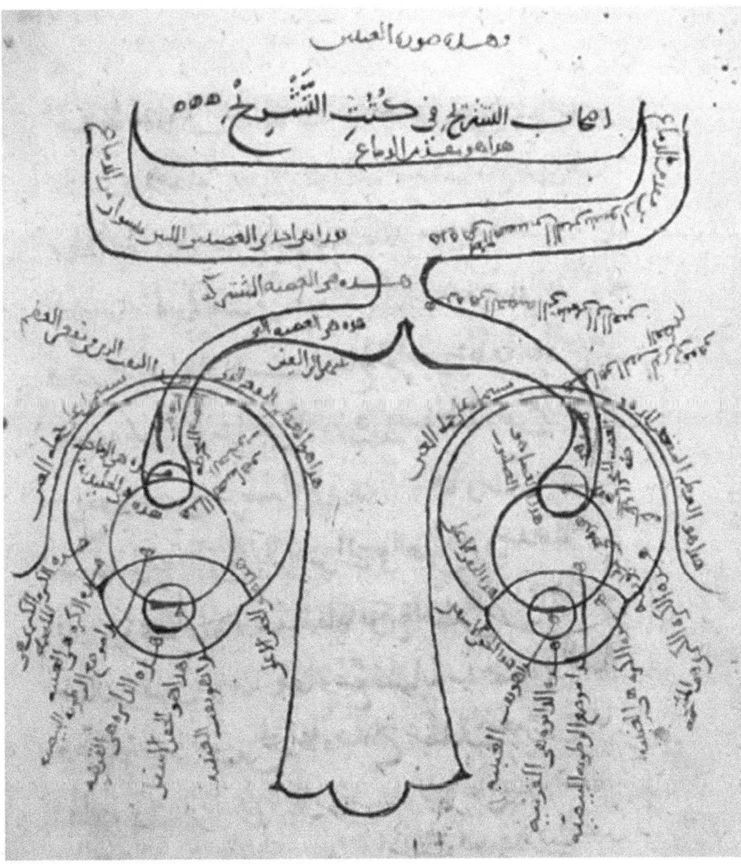

Figure 5.1 *Alhazen, Kitāb al-Manāẓir*, MS Fatih 3212, Vol.1, fol.81b, Süleymaniye Mosque Library, Istanbul.

Notes

1 Bacon, *Opus maius*, vol. II.
2 Ibn al-Haytham, *Kitāb al-manāẓir,* vol. I–III; Ibn al-Haytham, *Kitāb al-manāẓir,* vol. IV–V; Ibn al-Haytham, *The Optics, Books I–III*; Ibn al-Haytham, *The Optics, Books IV–V*; and Alhazen, *Opticae Thesaurus*. The Arabic title *Kitāb al-manāẓir* is usually translated in English as the *Book of Optics*; however, it can also literally mean the *Book of Perspectives*, *Book of Visible Aspects*, or *Book of Sceneries*.
3 My article is partly grounded on my earlier published studies on Ibn al-Haytham, such as: El-Bizri, "Alhazen's Phenomenology of Visual Perception," 202–218; El-Bizri, "The Confluences of Science, Philosophy, and Art," 65–68, 112–117; El-Bizri, "Ibn al-Haytham and Hevelius' *Selenographia*," 55–59; El-Bizri, "Alhazen's Theory of Vision," 21–33; El-Bizri, "Arabic Classical Traditions in the History of the Exact Sciences," 271–277; El-Bizri, "Desargues' oeuvres: On Perspective, Optics and Conics," 36–51; El-Bizri, "Seeing Reality in Perspective," 25–47; El-Bizri, "By Way of an Overture," 11–26; El-Bizri, "Classical Optics and the *Perspectiva* Traditions Leading to the Renaissance," 11–30; El-Bizri, "Ibn al-Haytham et le problème de la couleur," 201–226; El-Bizri, "In Defence of the Sovereignty of Philosophy," 57–80; El-Bizri, "A Philosophical Perspective on Alhazen's *Optics,*" 189–218; and El-Bizri, "La perception de la profondeur," 171–184.
4 The earliest printing of the Latin translation of Alhazen's *Optics* from Arabic is set in: Friedrich Risner, ed., *Opticae thesaurus*; an introduction was added by David Lindberg for the New York Johnson Reprint version of 1972. The documented reception and adaptive assimilation of Alhazen's *Optics* in the European scholarly milieu from the high Middle-Ages through the Renaissance and into the early-modern period passes via Gerard of Cremona's circle, Roger Bacon's *Opus maius*, John Pecham's *Perspectiva communis*, Erazmus Ciolek Witelo's *Opticae*, Theodoric of Freiberg's *De iride*, Biagio Pelacani da Parma's *Questiones super perspectiva communi*, Leon Battista Alberti's *De pictura*, Lorenzo Ghiberti's *Commentarii*, Friedrich Risner's, *Opticae thesaurus*, François d'Aguilon's *Opticorum,* Johannes Hevelius' *Selenographia*, Girard Desargues' *Manière universelle pour pratiquer la perspective*, René Descartes', *Dioptrique*; Johannes Kepler's, *Astronomiae pars optica.*
5 d'Aguilon, *Opticorum libri sex*.
6 al-Dīn al-Fārisī, MS Istanbul, Topkapı Kütüphanesi, Ahmet III 3340 (copied in Nīshāpūr, on 15 Sha'bān AH 716/1316 CE); edited in two volumes in Hyderabad (Deccan), 1928–1930.
7 Section 63 on "Late Medieval Optics" with a focus on Theodoric of Freiberg, *De iride*, 435–441.
8 Based on the tenth-century CE treatise of the *Kāmil al-ṣināʿa al-ṭibbīya* (*The Complete Medical Art*) or *Kitāb al-malakī (The Royal Book*; *Liber Regalis*) that was composed by Haly Abbas (ʿAlī ibn al-ʿAbbās al-Majūsī). See Burnett and Jacquart, *Constantine the African and ʿAlī ibn al-ʿAbbās al-Mağūsī.*
9 Black, "Estimation (*Wahm*) in Avicenna," 219–258. See also El-Bizri, "Avicenna and the Problem of Consciousness," 45–53.
10 See Leon Battista Alberti, *On Painting*; Ghiberti, *Commentarii*.
11 I investigated some of these aspects in architecture and the visual *cum* plastic arts in: El-Bizri, "The Confluences of Science, Philosophy, and Art"; El-Bizri, "Desargues' oeuvres: On Perspective, Optics and Conics"; El-Bizri, "Seeing Reality in Perspective"; El-Bizri, "By Way of an Overture: Classical Optics and Renaissance Pictorial Arts"; El-Bizri, "Classical Optics and the *Perspectiva* Traditions Leading to the Renaissance."

Bibliography

Manuscripts

al-Dīn al-Fārisī, Kamal. MS Istanbul, Topkapı Kütüphanesi, Ahmet III 3340 (copied in Nīshāpūr, on 15 Shaʿbān AH 716 / 1316 CE); edited in 2 volumes in Hyderabad (Deccan), 1928–1930.

Primary Sources

Francois d'Aguilon. *Opticorum Libri Sex*. Antwerp: Jan Moretus Widow & Sons, 1613.

Leon Battista Alberti. *On Painting*. Edited and translated by Rocco Sinisgalli. Cambridge: Cambridge University Press, 2011.

Lorenzo Ghiberti. *I Commentarii*. Edited by Lorenzo Bartoli. Florence: Giunti, Biblioteca Nazionale Centrale di Firenze, 1998.

Ibn al-Haytham. *Kitāb al-manāẓir*. Edited by Abdelhamid I. Sabra, vols. I–III. Kuwait: National Council for Culture, Arts, and Letters, 1983.

Ibn al-Haytham. *Kitāb al-manāẓir*. Edited by Abdelhamid I. Sabra, vols. IV–V. Kuwait: National Council for Culture, Arts, and Letters, 2002.

Ibn al-Haytham. *The Optics, Books I–III*. Translated by Abdelhamid I. Sabra. London: Warburg Institute, 1989.

Ibn al-Haytham (Alhazen). *Opticae Thesaurus*. Edited by Friedrich Risner. Basel: Episcopius, 1572. Latin critical edition and annotated English translation by A. M. Smith. Philadelphia: American Philosophical Society, 2001–2010.

Ibn al-Haytham. *The Optics, Books IV–V*. Translated by Abdelhamid I. Sabra and prepared for publication by Jan Hogendijk. London: Warburg Institute, 2023.

Roger Bacon. *Opus majus*. Edited by John Henry Bridges. *3 vols*. Oxford and Edinburgh: Clarendon, 1897-1900.

Theodoric of Freiberg. *De iride*. Translated by David Lindberg, in: *A Source Book in Medieval Science*, edited by Edward Grant, 435–441. Cambridge, MA: Harvard University Press, 1974.

Secondary Sources

Black, Deborah L. "Estimation (*Wahm*) in Avicenna: The Logical and Psychological Dimensions." *Dialogue: Canadian Philosophical Review* 32/2 (1993): 219–258.

Burnett, Charles and Danielle Jacquart. *Constantine the African and ʿAlī ibn al-ʿAbbās al Maǧūsī: The Pantegni and Related Texts*. Leiden: E. J. Brill, 1994.

El-Bizri, Nader. "Alhazen's Phenomenology of Visual Perception." In *The Book of Sceneries*, edited by Ala Younis and Tarek ElFetouh, 202–218. Berlin: Hatje Cantz Verlag, 2022.

El-Bizri, Nader. "Alhazen's Theory of Vision." *Micrologus* 29 (2021): 21–33.

El-Bizri, Nader. "Arabic Classical Traditions in the History of the Exact Sciences: The Case of Ibn al-Haytham." *The European Physical Journal Plus* 133 (2018): 271–277.

El-Bizri, Nader. "Avicenna and the Problem of Consciousness." In *Consciousness and the Great Philosophers*, edited by Stephen Leach and James Tartaglia, 45–53. London: Routledge, 2016.

El-Bizri, Nader. "By Way of an Overture: Classical Optics and Renaissance Pictorial Arts." In *Recto-Verso: Redefining the Sketchbook*, edited by Angela Bartram, Nader El-Bizri, and Douglas Gittens, 11–26. Aldershot: Ashgate, 2014.

El-Bizri, Nader. "Classical Optics and the *Perspectiva* Traditions Leading to the Renaissance." In *Renaissance Theories of Vision*, edited Charles H. Carman and John S. Hendrix, 11–30. Aldershot: Ashgate, 2010.

El-Bizri, Nader. "The Confluences of Science, Philosophy, and Art: Ibn al-Haytham's *Optics* and the European *Perspectiva* Traditions." In *From Wisdom to Data: Philosophical Atlas on Visual Representations of Knowledge*, edited José Higuera Rubio, Alberto Romele, Dario Rodighiero, and Celeste Pedro, 65–68, 112–117. Porto: University of Porto Press, 2022.

El-Bizri, Nader. "Desargues' oeuvres: On Perspective, Optics and Conics." In *Visioning Technologies in Architecture*, edited by Graham Cairns, 36–51. Aldershot: Ashgate, 2016.

El-Bizri, Nader. "In Defence of the Sovereignty of Philosophy: al-Baghdādī's Critique of Ibn al Haytham's Geometrisation of Place." *Arabic Sciences and Philosophy* 17 (2007): 57–80.

El-Bizri, Nader. "Ibn al-Haytham and Hevelius' *Selenographia*." In *The Mantis Shrimp: A Simon Schaffer Festschrift*, edited by Dániel Margócsy and Richard Staley, 55–59. Cambridge: The Cambridge History and Philosophy of Science Collective, 2022.

El-Bizri, Nader. "Ibn al-Haytham et le problème de la couleur." *Oriens-Occidens* 7 (2009): 201–226.

El-Bizri, Nader. "La perception de la profondeur: Alhazen, Berkeley et Merleau-Ponty." *Oriens Occidens* 5 (2004): 171–184.

El-Bizri, Nader. "A Philosophical Perspective on Alhazen's *Optics*." *Arabic Sciences and Philosophy* 15 (2005): 189–218.

El-Bizri, Nader. "Seeing Reality in Perspective." In *the Art of Science: From Perspective Drawing to Quantum Randomness*, edited by Rossella Lupacchini and Annarita Angelini, 25–47. Dordrecht: Springer, 2014.

6 Bacon on Mathematics, *Perspectiva*, and Astrology

The "Unnamed Master" Revisited

H Darrel Rutkin

6.1 Introduction

Astrology is a significant feature in a range of Roger Bacon's writings and in his vision of the world overall. I will focus here primarily on the *Opera tria* and related works. In this chapter, I will first describe how astrology was configured within his vision of the sciences, focusing, in particular, on his understanding of mathematics. In the second part and within this context, I will readdress the question of the so-called Unnamed Master in Bacon's *Opus tertium*, specifically regarding his views on mathematics and astrology in relation to *perspectiva*. In so doing, I will describe, in some detail, Bacon's vision of a deeply mathematicised, astrologising Aristotelian natural philosophy, with its geometrical-optical model of celestial influences within a Ptolemaic cosmographic framework.[1]

6.2 Astrology and Mathematics

6.2.1 Roger Bacon on Mathematics

I would like to begin this chapter by locating astrology within Bacon's broader map of knowledge, specifically in relation to mathematics. In the *Opus maius*, towards the beginning of Book IV, Bacon describes how he understands the place of mathematics within his broader map of knowledge, and some of its subdivisions. The first part of *Opus maius* IV, entitled "The Utility of Mathematics in Natural Philosophy" (97–174), is clearly structured. The passage of interest (109–111) directly follows a discussion *De scientiis* (98–108), where Bacon shows by authority (98–103) and by reason (103–108) why every science requires mathematics.

Bacon begins with a characteristically strong claim: "[I]t is impossible for the things of this world to be known, unless mathematics is known." He immediately begins justifying this universal claim by turning to the science of the stars:

> For the celestial [sc. bodies] (*coelestia*) are certain to everyone because there are two great mathematical sciences (*duae scientiae magnae mathematicae*) concerning them, namely *astrologia speculativa* and *astrologia*

DOI: 10.4324/9781003405702-7

> *practica*. The first contemplates [1] the quantities of all things that are in the heavens, and everything that is reduced to quantity, both discrete and continuous quantity. For it certifies the number of the heavens and of the stars, whose quantity can be grasped by instruments, and all of their figures, and their magnitudes and altitudes from the Earth, and their thickness and number, magnitude and smallness, the rising and setting of the signs of the stars, and the motion of both the heavens and the stars, and the quantities and varieties of eclipses. Then, it descends to [2] the quantity and shape of the habitable world, and of all its large parts, which are called *climata*, and it shows the diversity of horizons, and of days and nights corresponding to individual climates. These things, then, are determined here, and much is joined to them.[2]

In the first division of *astrologia speculativa* [1], Bacon straightforwardly describes the content and aim of what we call mathematical astronomy; in [2], he describes some of the basic divisions of mathematical geography.

Bacon then descends to the practical level:

> But [*astrologia*] *practica* descends to this, that [3] for every hour we would know the location of the planets and stars, and their aspects and relative positions and everything which is periodic among celestial bodies, and it descends to [4] those things that come to be in the air, of which sort are comets, and rainbows, and other periodic things there, in order that we know their locations, and altitudes, and magnitudes, and shapes, and many things which it ought to consider in these matters. And these all come to be through instruments fitted to these ends, and by tables, and by canons, that is, rules devised for confirming these things.[3]

Let us stop here momentarily to note that this first part of practical *astrologia* concerns the knowledge of where the planets and stars are at every hour of every day, and of certain meteorological phenomena, and that this knowledge is attained with the help of certain unspecified instruments. We are thus still well within the realm of astronomy proper, but as particularised for specific times.

The second part of *Astrologia practica* continues thus:

> [S]o that the way is prepared [5] for judgements [sc. or interpretations] (*ad iudicia*), which can be made in accordance with the power of philosophy, not only in the realm of natural things, but in those which take their inclination from nature, and freely follow the celestial disposition. Not only is the way prepared for judgements (*iudicia*) of the present, past and future, but for amazing works (*ad opera miranda*), in order that everything favourable for this world may be promoted, and what is adverse repressed, usefully and magnificently. These things are not in

> doubt. For the patriarchs and the prophets have confirmed these and other things from the beginning of the world. And Aristotle renewed the confirmation of the ancients, and brought it into the light. And all wise men agree in this in great matters, and experience teaches it. There will be an exposition of these things in their place.[4]

Nec sunt haec dubia, Bacon proclaims, adducing the support of authority – both religious and secular – and experience.

6.2.2 Iudiciaria-Operativa *Distinction*

After showing how Bacon configured astrology within the science of the stars, and how he distinguishes what we call "astronomy" from what we call "astrology," both of which he calls *astrologia* in this programmatic passage, I will now indicate how he further articulates what we call astrology itself, to which he had just alluded. Here, I will treat his explicit configuration of astrology within the mathematical disciplines in the introduction to his important and slightly later edition of the *Secretum secretorum*, and especially its bifurcation into [1] an interpretive knowledge-based part – which he calls *iudiciaria* – and [2] an operative or what we would call magical or technological part – which he calls *operativa* – that is concerned with acting in the world.

After introducing the distinction between true and false mathematics within the context of divination and astrology, a topic to which we shall return shortly, Bacon discusses astrology's configuration among the "true" mathematical disciplines: "And this *mathematica* contains four sciences (*scientiae*), namely geometry, arithmetic, music and *astrologia*, under which *astrologia*, *astronomia iudiciaria* and *operativa* are contained in a common term."[5] We can see clearly here that the four *scientiae* contained within true mathematics are the four normal disciplines of the *quadrivium*, namely, geometry, arithmetic, music, and the science of the stars.

More interesting for our purposes is that, after normatively configuring the science of the stars within the mathematical disciplines, Bacon further distinguishes *astrologia* into two parts. He refers to the first as *astronomia iudiciaria,* which I will neutrally and loosely translate here as "the judicial or interpretive part of the science of the stars," and the second as *astronomia operativa* or "the operative part of the science of the stars." As it turns out, the latter is concerned with two different practices that Bacon refers to as the works and words of wisdom (the *opera et verba sapientiae*), namely, making works (*opera*) – primarily talismans – or uttering potent and powerful words (*verba*) in order to move or affect things in the world. We would normally call both of these "magical" practices, but Bacon himself most certainly would not do so, as we will see in a moment. Bacon's distinction here is extremely useful for understanding the medieval map of knowledge regarding what *we* call both astrology and magic. Although this phrasing appears in

Bacon's introduction to his slightly later edition of the *Secretum secretorum*, it concisely and precisely captures structures he expounds at much greater length in the three *Opera* composed for Pope Clement IV.

With these basic conceptual structures in mind, let us now turn to Bacon's defence of astrological interpretations or judgements (*iudicia astronomiae*) in *Opus maius* IV. My focus here will primarily be on the *iudiciaria* side of the *iudiciaria-operativa* distinction, although the operative side is also very interesting, especially the making of talismans, which is covered in greater detail in Nicolas Weill-Parot's chapter in this volume. Bacon's usage here can also help us to more properly historicise the more modern natural-judicial astrology distinction, which is nowhere apparent in Bacon's work, and should thus not be imposed thereon, at least not without good reason.

6.2.3 *Roger Bacon's Defence of Astrology: True vs. False Mathematics*

In *Opus maius* IV (238–267), Bacon begins a new section that is centrally concerned, first, with removing all doubts and concerns about mathematics, and ultimately with establishing its utility for theology. Given the centrality of mathematics to his overall natural philosophy, and of philosophy's centrality to theology, Bacon devotes significant attention to defending mathematics' weakest and most controversial link, namely, what he calls the judgements of the science of the stars (*iudicia astronomiae*), which we normally call astrology. Bacon's attention to these issues is fully justified within his system of thought, and it is very interesting for our purposes.

To defend astrological judgements or interpretations, Bacon begins by introducing an essential, characteristic, and utterly fundamental distinction for him between true and false mathematics (his terminology), which leads into a rich discussion of both astrology and what *he* explicitly calls "magic":

> Having made clear how *mathematica* is necessary for both divine and human wisdom, it is also necessary for certifying what went before, namely, that certain complaints against it [sc. mathematics] be removed, and that certain sayings of the saints be set out, so that every doubt (*omnis dubitatio*) is destroyed concerning the utility of mathematics. That respect in which mathematics is primarily attacked is due to astrological judgements (*iudicia astronomiae*).[6]

Bacon immediately identifies mathematics' weakest link – and thereby the source of its attacks – as what he calls *iudicia astronomiae*.

Bacon begins his defence by attacking ignorant theologians, who, in his view, created the problem in the first place due to their systematic misunderstandings:

> Therefore, since the contention of the many people ignorant of the power of philosophy and its great utility for theology, and the considerations

> of those who reprove the *mathematici* – both relatively and absolutely – impede the study of wisdom and gravely harm it in this part, therefore, I wish at present to verify their intention, and remove their disrespect (*infamia*) from true mathematics (*vera mathematica*).[7]

Because mathematics has power for philosophy and great utility for theology, it must be defended, and its bad reputation removed by a process of reconceptualisation.

After putting forward the concept of true mathematics, Bacon immediately and explicitly distinguishes it from false mathematics. He first discusses the statements of the saints and offers his interpretation of their intentions:

> Therefore, theologians have found many things uttered by the saints against the *mathematici*, and some of them [sc. the theologians] – because of their ignorance of true mathematics and false mathematics – do not know how to distinguish true from false, and therefore, as though by the authority of the saints, they blame true mathematics with [sc. the errors of] the false.[8]

Due to their ignorance of the true nature of mathematics, certain theologians have confused true with false mathematics and thus mistakenly attacked the true kind by conflating the two, thereby reproaching both.

To set the record straight by properly distinguishing true from false mathematics, Bacon introduces an etymological argument on which he places great weight, but which changed diametrically over time:

> For the word (*vocabulum*) for true mathematics is written with an aspirated "t", and from this, the noun (*nomen*), *mathesis*, with the middle shortened (*correpta*), [sc. is derived], which signifies knowledge (*designat scientiam*). It is said to be derived from many authors [or authorities, *auctores*]. It is certainly from Greek, since "*matheo*" is the same verb (*verbum*) as "*disco*" [= I teach] is, and "*mathetes*" is "*discipulus*" [= student], and "*mathesis*", "*disciplina*" [= teaching, body of knowledge]. Whence *mathematica* is "*disciplinalis scientia et doctrinalis*" as Cassiodorus said. But the word (*vocabulum*) for false mathematics is written without an aspiration and is asserted by the same authorities, and is from *mathesis* with the middle lengthened (*media producta* [vs. *correpta*]), which denotes divination (*divinatio*) and derives [sc. etymologically] either from "*mantos*" (which is more certain) or from "*mantia*", which are both the same as "*divinatio*", as Jerome says in the original of IX Isaiah.[9]

Although the terminology for true and false mathematics is distinctive, it is also deceptively similar, boiling down to the difference between a "t" vs a "th" sound and a shortened or lengthened vowel quality with correspondingly different etymologies. True mathematics is thus derived from *matheo* and *mathesis,* and is pronounced with an aspirated "t" and a shortened vowel quality. It means "learning" and "knowledge." False mathematics, on the other hand, derives from *mantos* or *mantia* and refers to divination. A further distinction here between true and false *knowledge* is also implied. Thus, divination as false knowledge or learning is decidedly negative for Bacon, which was also the case for Thomas Aquinas, as I discuss elsewhere.[10]

As is well-known, Bacon subsequently revised his analysis of this same etymology in his edition of the *Secretum secretorum*, which he wrote at some point between 1270 and 1292.[11] Despite the switched terminology with its correlative etymologies, however, the results of his later analysis were exactly the same, in that Bacon still conceptually distinguished true from false mathematics in precisely the same manner. The weight of his concern thus falls much more heavily on the conceptual distinction than on the terminology with its respective etymologies.

To complete this brief initial etymological discussion, Bacon makes another major point that he will return to often. In addition to divination (or false knowledge), false mathematics is also closely associated with what he explicitly calls the magical art. In Bacon's usage, what he explicitly calls magic and divination are *always* negative:

> But whatever the case is about this writing (*scriptura*, spelling) and its [sc. etymological] derivation, false mathematics is, nevertheless, a magical art (*ars magica*). For five species of the magical art are enumerated, namely, *mantice* [= divination], *mathematica* [= mathematics or astrology], *maleficium* [= doing harm], *praestigium* [= illusions] and *sortilegium* [= the casting of lots (sometimes translated as "sorcery")]. Therefore, false mathematics is the second part of the magical art. It usurps to itself a consideration of the celestial bodies deformed by characters, charms, conjurations, superstitious sacrifices and various frauds.[12]

Bacon here offers a fivefold typology of the magical arts, adopting the structure from Hugh of St. Victor's *Didascalicon*. False mathematics is thus included integrally among the magical arts, as is *mantice* or divination. They are all anathema, as Bacon proceeds to discuss. Astrology, magic, and divination are thus all closely associated here, and we can begin to see how they are configured within Bacon's thought with respect to both his distinctive terminology and its underlying conceptual structures.

Bacon then strongly asserts why this false magical mathematics is problematic, and in so doing, indicates the main issues for a theologian:

> And it [sc. mathematical magic, or magical and thus illegitimate astrology] posits [1] that everything happens necessarily (*omnia de necessitate contingere*) by the power of celestial configurations (*per virtutem*

> *constellationum*), [2] that nothing can go one way or another (*nihil ad utrumlibet*), [3] that nothing happens by chance or fortune (*nihil a casu nec fortuna*), and [4] nothing by counsel (*nihil a consilio*)[.][13]

This false type of mathematics – that is, illegitimate magical astrology – both assumes and implies a deterministic structure of nature, in which everything happens by necessity in such a way that free will, chance, and human choice are wholly undermined. The terminology Bacon uses is significant. The implications of this view for free will in particular, both divine and human, had long made astrology suspect. Bacon will discuss this subject at length in what follows in *Opus maius* IV and return to it repeatedly as a leitmotif.[14]

Bacon completes this section by relating it back to and thus concluding the larger argument being addressed here, namely, defending mathematics' value in relation to its dubious associations by some with astrological judgements. To do so, he defends true mathematics in the central and controversial realm of both individual and collective human affairs, where concerns about the implications of necessity in nature and infallible astrological judgements are both successfully addressed. In the process, Bacon provides us with valuable insight into his views on the roles of nativities and revolutions – two of the four canonical types of practical astrology – with both medical and political applications, and thus of practical astrology's profound utility in human affairs.

As it turns out, Bacon also uses astrological revolutions in his analysis of the history – i.e., the generation and corruption – of the great world religions in terms of both the regularly recurring great conjunctions of Jupiter and Saturn and a series of ten Saturn cycles.[15] He primarily discusses elections – the choosing of astrologically propitious times – in relation to making talismans, the *opera sapientiae*.[16] As far as I know, however, Bacon does not use interrogations, the last of the four canonical types of astrological practice.[17]

6.3 The "Unnamed Master" Revisited

Now that we have some idea of Roger Bacon's views on astrology's central place within his map of knowledge, I would like to re-approach the question of identifying the "Unnamed Master" in Bacon's *Opus tertium* within this context. I will approach this reappraisal by asking a central question, namely, "What is the intellectual relationship between two of the most important thirteenth-century natural philosophers, Albert the Great and Roger Bacon?" More specifically, is Albert the unidentified object of Bacon's spleen in the *Opus tertium*? The *status quaestionis* currently leans quite heavily in the affirmative. The two most recent scholars to discuss this issue in any depth, Stewart Easton in 1952 and Jeremiah Hackett in 1980, both conclude that Albert is, in fact, the Unnamed Master; David Lindberg, Paola Zambelli, Amanda Power in her recent biography, and now Nikolaus Egel in his new edition of the *Opus tertium* push the scales even further in this direction.[18]

Nevertheless – and however plausible the attribution – we should keep the question open and take an approach that seems strangely neglected. Easton stated his desire to compare what Bacon said about the Unnamed Master with the known facts of Albert's life and works, and this is surely the soundest approach. In practice, however, Easton focused almost exclusively on theological issues, dismissing the evidence of Albert's scientific writings on his final page. Hackett, on the other hand, did not compare the content of their ideas at all. He was content, rather, to show that the named attributions to Albert in Bacon's works were similar to what was said of the Unnamed Master, and that, therefore, the Unnamed Master is Albert. Even if it were otherwise compelling, however, Hackett undercuts this argument on the final page of his essay, where he explicitly states that proper names in Bacon's manuscripts are not particularly reliable.

Fortunately, Bacon's criticisms of the Unnamed Master are detailed enough to reconstruct his views on the subjects in question. Likewise, there is much relevant material by Albert. In fact, the content of Bacon's criticism, on which we shall focus—the Unnamed Master's ignorance of *perspectiva* – provides an excellent basis for comparison.

Bacon mentions the "Unnamed Master" in his *Opus tertium*, the overall prolegomenon to his set of three works, addressed and sent to Pope Clement IV in 1266–1267, which has recently received a valuable new critical edition of the Latin text by Nikolaus Egel.[19] The main passages of criticism occur in Chapter XI, where Bacon discusses first mathematics and then *perspectiva*, explicitly criticising the Unnamed Master twice for his utter ignorance of *perspectiva*. Here is the first criticism:

> And it is necessary that everything be known by this science [sc. *perspectiva*] because all the actions of things come to be in accordance with the multiplication of species and powers, from the agents of this world [sc. acting] on receptive matter. And the laws of multiplications of this sort are not known except by *perspectiva* (and they are not treated anywhere else), since not only are actions common with regard to vision, but they are so for all the senses, and in the entire system of the world, both in the heavens and on Earth. But this science has not yet been taught in Paris, nor among the Latins, except twice at Oxford in England; and there are not three people who know its power. Therefore, he who made himself an authority (an *'auctor'* [the first explicit criticism]), about whom I spoke above, knew nothing of the power of this science, as appears in his books, because he did not compose a work on this science—he would have done so had he known it—nor does he say anything on this science in his other works.[20]

For our purposes, we should first note the somewhat detailed discussion of *perspectiva* encountered before the criticism, and secondly, the clearly stated claims that the Master, whoever he is, never wrote a treatise on *perspectiva*, nor did he treat it at all in any of his other works. Since we do not, in fact, possess a separate treatise on *perspectiva* by Albert, the attribution is supported – so its accuracy will stand or fall on the second point: if, in fact, there are no discussions at all of *perspectiva*, as understood by Bacon, in Albert's extant works, then we shall be able to lend solid, albeit negative, evidence to his attribution as the Unnamed Master.

Bacon then provides references to his fuller treatments on *perspectiva,* namely, *Opus maius*, Book V, a full treatise on *perspectiva*, and a large part of Book IV, where he "determined on the multiplication of species and the virtues of agents," which he "took from the science of *perspectiva*":[21]

> And for this reason, because it is the highest and principal root of wisdom, both for philosophy and for theology, there is in these multiplications also unbounded beauty; and neither is *perspectiva* nor any bit of philosophy able to be known without it. But he—[and here we have the second criticism]—who multiplied volumes, ignores these roots, for he touches on none of them. And therefore it is certain that he is ignorant of natural things, and everything concerning philosophy; and not only him, but the whole horde of philosophizers who err because of him.[22]

Albert certainly did write a lot, as any glance at his collected works reveals, so the comment about the Unnamed Master "multiplying volumes" could surely apply to him as well; likewise, could the comment about his great influence on students of philosophy. There can be no doubt of the plausibility, in general, of thinking that the Unnamed Master is Albert. This conclusion is further supported by the lack of other compelling candidates.

Bacon discusses his views on *perspectiva* in greater detail in *Opus maius,* Book IV, and even more so in Book V. I will focus on his concern with places in the world and their importance in generation in Book IV. He states the main issue most clearly in a section entitled "Geography":

> For the things of the world cannot be known except through knowledge of the places in which they are contained. For place is a principle of the generation of things, as Porphyry says, because the diversity of things corresponds to the diversity of places, and not only of natural things, but also of moral and scientific [....] We can verify this based on understanding the longitude and latitude of any place. For then we know under which stars each place is, and how far it is from the path of the sun and planets, and of which planets and signs the places receive dominion, which all make the diverse complexions of places. If they were known, one could know the complexions of every thing in the world, and the natures and properties which they draw from the *virtus* of a place.[23]

Here, Bacon implicitly refers to his earlier discussion of *astrologia speculativa.*

Let us begin to unpack what Bacon says here by turning to a section where he discusses the problem *De natura loci*, on the nature of place:

> [C]omplexions of the places of the world cannot be discovered unless a person knows the aforementioned laws of multiplications [sc. of species] [....] [P]lace is a principle of generation, as also a father is, as Porphyry says (a). And we see (b) that everything is varied in accordance with the diverse places of the world.[24]

Bacon then makes the picture more precise by providing the causal basis of his theory:

> But since the things of this world which have been established in diverse places, however close they are, receive cones of diverse pyramids coming from the entire heavens opposite them, therefore unbounded diversity arises. For the cones of individual pyramids come to single points of the Earth, and each point [sc. of the Earth] is the centre of a new horizon.[25]

Here Bacon begins to develop an analysis whereby natural things cannot be known except by their causes, namely, the heavens – luminaries and planets within the zodiac – which he places into a richly mathematised scientific framework informed by *perspectiva*. Here he establishes the geometrical uniqueness of each point on Earth based on its unique angular relationship to the celestial situation, which is a central feature of his system.

Bacon then illustrates the significant implications of his analysis:

> And therefore we see that two plants born at the same time from the Earth without anything in between them, and therefore two twins in the womb of their mother, are allotted a difference of complexion, so that afterward they would have diverse customs, and they would pursue different arts and different occupations throughout their entire lives. And therefore the powers of the heavens and the stars produce everywhere things diverse in properties and natures, that is, in things generated by propagation [....] And when a child at birth is exposed to a new air as to another world, then he receives cones of celestial pyramids with respect to his individual parts, and thus he receives new impressions which never leave, because "what a new [= young] head grasps, it knows forever." And then the root complexion is confirmed, which always remains until the end of life, although the current complexion changes with every day. And it is this root complexion which the inclinations to customs and to sciences and to languages follow, and to whatever handiworks and occupations, and to every difference that we see in everyone.[26]

In this section, Bacon shows the necessity of his doctrine, the multiplication of species, for understanding the complexions of things in this world, namely, their characteristic qualitative natures. Celestial influences are fundamental, directly affecting both levels of a complexion, the root and the current. He also thereby provides the natural philosophical foundations for nativities or natal astrology.

Now that we have a firmer grasp of Bacon's views on *perspectiva* – and their bearing on the criticism of the Unnamed Master – I will now more systematically describe the central feature of his model of action, what David C. Lindberg calls "pyramids of force." Bacon emphasises the importance of this pyramidal figure:[27]

> And this is the figure which nature especially chooses in every multiplication and action, and not any pyramid, but that whose base is the surface of the *agens*, and whose cone falls on some point on the *patiens*, because thus the species can come from the entire surface of the *agens* to the individual points of the *patiens* through an infinite number of individual pyramids[.][28]

He then describes the structure of how the *agens* acts:

> It is fitting that the multiplication comes to be spherically. For the *agens* multiplies itself equally on every side, and in accordance with every diameter and every difference of position, which are up, down, fore, aft, right and left. Therefore, lines come forth from everywhere on all sides from an *agens*, just as from the centre.[29]

This omnidirectional radiation, the multiplication of an agent's species, is analysed in terms of these pyramids for every point where its action is received.

Once these pyramids of force are understood in the abstract, Bacon can then show how his model accounts for celestial influences. In order to analyse accurately the pyramid of force resulting from any given celestial configuration, each of the planets must first be accurately located. Once a working theory of each planet's motion has been devised, and hopefully verified by observation, then, for convenience, tables may be constructed for their location at every hour of every day, precisely the aim of *astrologia practica* as discussed above. This is critical because every planet, as an *agens* in Bacon's model, continuously emits an infinite number of rays in every direction. If each planet is accurately located at every time, its unique angular relationship to the centre of the Earth can then be known.

One must then be able to localise these planetary patterns of influence for particular places at particular times. In order to do so, the mathematical-geographical framework must be added by accurately mapping the longitudes and latitudes of the Earth and then calibrating them with the patterns established by mathematical astronomy. Then, one may localise celestial influences within this essentially Ptolemaic cosmographic framework by means of the horizon, which is defined precisely – and uniquely for each place – by the relation of each point on the sphere of the Earth to its correspondingly unique point on the heavenly sphere. Bacon concludes:

> [T]o every point of the Earth falls the cone of one pyramid made virtuous by the entire heaven. And these cones are diverse in nature, and the pyramids likewise, because they have different bases because of the differences of their horizons, since each point of the Earth is the centre of its own horizon.[30]

It is only at this point, when the integrated mathematical, astronomical, and geographic framework has been constructed – namely, when the abstract pyramids of force can be instantiated for a particular planetary configuration at a particular time for a particular place – that, and only then, can the astrologising natural philosopher approach his or her proper questions.

The basic fourfold structure of planetary influences corresponds to these four questions: first, what is the qualitative nature of each planet? Second, what is the qualitative nature of each planet as modified by each sign of the zodiac? Third, what, then, is the qualitative nature of the figure of the heavens at any given time, that is, what is the qualitative relationship of each planet in each of their respective signs at a particular time *in relation to each other*? Finally, what is the qualitative nature of the overall celestial configuration in its unique angular relationship to each particular point on the sphere of the Earth?

The fourth question relates directly to Bacon's concern with place and thus fits squarely into the causal framework articulated here. The figure of the heavens at all times is related to every place in the world here below, irradiating it with its particular mix of planetary qualities at that time, and thus influencing and thereby informing its complexion. Then, since place is a principle of generation, this planetary configuration affects the complexion of every natural thing generated there. This omnidirectional irradiating celestial configuration is described by a complex pattern of planetary pyramids of force, which collectively have a unique angular relationship to every point on Earth. This last angular relationship is important because whether the angles of incidence are perpendicular or oblique further conditions their strength.

As I have now indicated, Roger Bacon is remarkably consistent in his analysis of the variety of places in the world and their fundamental importance for understanding the complexions of naturally generated things. He uses the same causal analysis throughout, namely, the multiplication of species, a geometrical-optical model of action, planetary and otherwise, within a fundamentally Ptolemaic cosmographic framework composed of mathematical astronomy integrated with mathematical geography, that is, the very themes which informed his detailed criticism of the Unnamed Master.

6.3.1 *Albert's Position*

We are now ready to compare Bacon's views with those of Albert. I will discuss an obvious work, a treatise entitled *On the Nature of Place* (*De natura loci*). It falls within the series of his Aristotelian commentaries, filling a lacuna in Aristotle's system where no authentic work exists. Albert places it just after the *Physics, De generatione et corruptione,* and *De caelo*, thus among Aristotle's core natural philosophical works, and it may be dated to circa 1251–1254, while he was at Cologne.[31]

Our first passage is the opening paragraph of the first chapter, entitled, "That It Is Proper in Natural *Scientia* to Know the Nature of Place":

> We will treat the nature of places, which comes from the relation of the place to the heavens; first we will mention those things determined in the *Physics* [IV.1.10]. For there it was proved that place is an active principle of generation, like the father.[32]

We thus immediately encounter the familiar view that the nature of place concerns the relationship of the heavens to the Earth and that place is a principle of generation.

In a second passage, we get more detailed information:

> Therefore, from all of these things, it is sufficiently allowed that one ought to know the nature of place. Nor does the treatise in the *Physics* suffice, for this reason, that it only verifies concerning place universally [....] Because of which, those err who treat *scientia naturalis* and do not bring in anything about the diversity of places. They seek solace for their inexperience if they say that it is not fitting to inquire about this, for it does not suffice that this sort of diversity be treated in mathematical [sc. books], where which stars rise or fall in this or that manner above which places is treated, namely, whether obliquely or [sc. on the perpendicular]. And which of them neither rise nor set, but are always apparent, or which never appear, and which rise and set sometimes. Because, although from these the location of the *climata* may be known, nevertheless the nature of places and the diversity of their complexions are not sufficiently known by means of these.[33]

Albert here indicates the necessity of mathematical astronomy and geography, but also their insufficiency in themselves for the natural philosopher. Bells of recognition will have gone off at the last phrase: that the nature of places and their diversity of complexion will not thus be sufficiently known.

In Chapter 5, Albert discusses the particularities of places. This passage provides material for a proper comparison:

> Moreover, if anyone wished to understand in particular all the natures and properties of particular places in water and air and on Earth, he

> would know that there is not a point in them which does not have a special property from the power of the stars oriented toward the middle habitation of mixed things [sc. the Earth], for the circle of the horizon is varied in relation to each point of the habitation of animals and plants and stones, and the entire orientation of the heavens to the middle place of habitation [sc. the Earth] is varied in relation to the variation of the horizon's circle. From which cause, their natures, properties, customs, actions and species, which seem to be generated in the same perceptible place, are varied, to such an extent that diverse properties and customs are attributed to twins' seeds, both for brute animals and for men, from this different orientation. And this is reasonable because it has been learned that the heavens pour forth formative powers into everything that exists. Moreover, it mostly pours them forth by means of rays emitted by the lights of the stars (*per radios emissos a luminibus stellarum*), and therefore it follows that each pattern and angle of rays causes different powers in things below (*quaelibet figura radiorum et angulus diversas in inferioribus causet virtutes*).[34]

There can be no doubt that Albert just provided exactly the same causal analysis as Bacon did, and in much the same language. In fact, Albert has now introduced a geometrical-optical model of celestial influences, where the different patterns of celestial rays striking different points on Earth cause different virtues in things below. He also uses the horizon to individuate the location of each and every place on Earth in relation to the celestial situation, just as Bacon did, and for the very same reasons.

The passage continues:

> Since, therefore, [1] from a change of the horizon, it is necessary that [a] the entire circle would be changed, and from the change of the circle, [b] the entire configuration of the rays would be changed (*tota mutetur figuratio radiorum*). And, since [2] each point of a habitation constitutes one special centre of a horizon, it follows necessarily that each point of the habitation has special powers, by which that which is located in it is informed. And this is the cause why no generated thing is found to be completely similar to something else in every detail of its properties. And this is *locus* properly, about which Porphyry says that it is one of the individuating features [....] A celestial configuration causes the qualitative power of the stars which are poured into things below (*[C]onstellatio est causans virtutem qualitativam stellarum, quae infunditur inferioribus*) [....] Moreover, not only is each particular place of a different property than another, but also each place in itself is made of different properties successively [sc. at different times], for this reason, that by means of the rise and fall of the stars above it [sc. the place in question], and the different relationship of these stars to each place, it

> is necessary for its properties to be varied. And this is the cause why the generated things are transformed by what contains them[.][35]

Albert's causal analysis is thus exactly the same as Bacon's, emphasising each place's unique qualitative makeup and that the nature of the celestial influences changes over time.

One last passage:

> Because of which, also, those who are wise concerning the stars first diligently consider the [sc. celestial] circle of each place, then they proceed to consider the powers of the places as a second [sc. set of] stars, and from these two joined together, they prognosticate about the nature of the thing generated (*prognosticare de natura rei generatae*). For the stars flow their powers in by means of something else, not without an intermediary, and they use a twofold medium for the influx of their powers, of which one is a ray and the second, a containing place. And by means of these two, they [sc. the wise astrologers] come to the matter of the generated thing, which the flowed-in virtues form and determine. Moreover, the variation of the stars' location is the cause of the variation of the figure (*Variatio autem situs stellarum est causa variationis figurae*) which the rays describe above the circle of the horizon (*quam describunt radii super circulum horizontis*). And this is the cause of the variation of the powers of places and of those things which are generated in places, and therefore it is necessary that a natural philosopher know these variations.[36]

6.4 Conclusion

Given this comparison, I think we must conclude that the contents of Albert's astrologising Aristotelian natural philosophy discussed here could hardly be closer to Bacon's, both conceptually and linguistically. Albert uses a geometrical-optical model of celestial influences within a fundamentally Ptolemaic mathematical, astronomical, and geographical framework to explain the fundamental role of place in generation, precisely as Bacon did. Perhaps we should conclude more accurately that central patterns in Roger Bacon's astrologising Aristotelian natural philosophy bear an extraordinarily close resemblance to Albert's, considering that Albert wrote his relevant work over ten years prior to Roger.

One might even think that Bacon completely took over the substance of Albert's work and simply refined it by adding Robert Grosseteste's notion of the multiplication of species. If Albert knew Grosseteste's work, he apparently decided not to embrace this characteristic feature of his analysis. In that case, Bacon seems to be asserting here, in effect, that ignorance of *perspectiva* can be claimed for simply not using the language of the multiplication of species because this seems to be the only discernible difference between Albert the Great and Roger Bacon's richly astrological analyses of place.

We seem to have two options: either Bacon intended the criticism to apply to Albert, or he did not. If he did, it is ill-founded, based on ignorance, bloody-mindedness, or both. If he did not, to whom was it meant to apply? Both options are problematic. The second would be better for Bacon's scholarly reputation but would leave a gaping hole in our knowledge. Other things being equal, we should choose it by the principle of interpretive charity – but where we can check other claims of Bacon against independent sources, we often find significant evidence of his equally ignorant and/or bloody-minded assertions, as in his well-known comments about Grosseteste and William of Moerbeke. Besides supporting the view that Bacon meant the criticism for Albert, the positive weight of this evidence also, happily, requires no search for other suitable candidates. Further, the unfortunate impact of this evidence on Bacon's already tarnished character is minimal, simply confirming what was already known.

We may conclude, then, that Bacon's criticism was probably intended to apply to Albert, even though it does not, in fact, apply to him, as we can judge from the independent evidence of Albert's scientific works, thus complicating the issue rather than settling it. Regardless, we now have a deeper understanding of the strikingly similar, intricately mathematicised, astrologising Aristotelianisms of two of the most important and influential thirteenth-century natural philosophers.

Notes

1 Rutkin, *Saptientia astrologica, ca. 1250–1800.* I must leave out much more on Bacon's astrology than I can discuss here in a chapter of reasonable length. Those who would like more detail and a much richer bibliography should refer to volume I of my monograph, *Sapientia astrologica*, about one quarter of which is devoted to Bacon. I make the argument addressed here concerning the "Unnamed Master" in greater detail in Chapter 2 of my volume I, which also provides a much richer bibliography. All translations are mine unless otherwise indicated. I would like to acknowledge that this article was conceived, composed, and completed as part of a project that has received funding from the European Union's Horizon 2020 Research and Innovation Programme (GA n. 725883 Early Modern Cosmology).

2 Bacon, *Opus maius*, vol. I, 109. De coelestibus enim certum est omnibus, quia duae scientiae magnae mathematicae sunt de eis, scilicet astrologia speculativa, et astrologia practica. Prima speculatur quantitates omnium quae sunt in coelestibus, et omnia quae ad quantitatem reducuntur, tam discretam quam continuam quantitatem. Nam numerum coelorum et stellarum, quarum quantitas potest per instrumenta comprehendi, certificat, et figuras omnium, et magnitudines et altitudines a terra ac spissitudines et numerum ac magnitudinem ac parvitatem, ortum et occasum signorum stellarum, et motum tam coelorum quam stellarum, et quantitates et varietates eclipsium. Item descendit ad quantitatem et figuram habitabilis, et omnium partium eius magnarum, quae vocantur climata, et ostendit diversitatem horizontium et dierum ac noctium secundum singula climata. Haec ergo determinantur hic, et multa eis annexa. See also the chapters by Jeremiah Hackett, Alexander Fidora, and Aurelien Robert in this volume.

3 Bacon, *Opus maius*, vol. I, 109. Practica vero descendit ad hoc, ut ad omnem horam sciamus loca planetarum et stellarum, et aspectus et compositiones earum

et omnia quae in coelestibus renovantur, atque descendit ad ea quae fiunt in aere, cuiusmodi sunt cometae, et irides, et caetera ibi renovata, ut sciamus loca eorum, et altitudines, et magnitudines, et figuras, et multa quae oportet considerare in his. Et haec omnia fiunt per instrumenta ad haec idonea, et per tabulas, et per canones, id est, regulas ad haec certificanda inventas.

4 Bacon, *Opus maius*, vol. I, 109-110. [Q]uatenus via paretur ad iudicia, quae fieri possunt secundum potestatem philosophiae, non solum in naturalibus, sed in his quae sumunt inclinationem ex natura, et gratis sequuntur coelestem dispositionem; et non solum ad iudicia praesentium praeteritorum et futurorum, sed ad opera miranda, ut omnia prospera huius mundi promoveantur, et adversa reprimantur, utiliter ac magnifice. Nec sunt haec dubia. Nam patriarchae et prophetae a principio mundi certificaverunt haec, sicut caetera. Et Aristoteles renovavit certificationem antiquorum, et produxit in lucem. Et omnes sapientes in rebus magnis in hoc concordant, et experientia docet. Sed de his expositio fiet suo loco.

5 Bacon, *Secretum secretorum*, 3. Et hec mathematica continet quatuor sciencias, scilicet, Geometriam, Arismetricam, Musicam, Astrologiam sub qua Astrologia, Astronomia iudiciaria et operativa vulgari nomine continetur.

6 Bacon, *Opus maius*, vol. I, 238–239. Manifestato quomodo mathematica necessaria est sapientiae tam divinae quam humanae, adhuc necesse est ad certificationem praecedentium, ut evacuentur quaedam cavillationes in contrarium et exponantur quaedam dicta sanctorum, ut omnis dubitatio tollatur circa mathamaticae utilitatem. Et illud in quo maxime percutitur mathematica est propter iudicia astronomiae.

7 Bacon, *Opus maius*, vol. I, 239. Quoniam igitur contentio multorum ignorantium philosophiae potestatem et maximam utilitatem theologiae, tam relate quam absolute reprobantium mathematicorum considerationes, impedit studium sapientiae et laedit gravissime in hac parte; ideo volo in praesenti verificare intentionem eorum, et evacuare infamiam verae mathematicae.

8 Bacon, *Opus maius*, vol. I, 239. Theologi igitur multa invenerunt a sanctis effusa contra mathematicos, et aliqui eorum propter ignorantiam mathematicae verae et mathematicae falsae nesciunt distinguere veram a falsa, et ideo tanquam auctoritate sanctorum culpant veram cum falsa.

9 Bacon, *Opus maius*, vol. I, 239. Vocabulum enim verae mathematicae scribi per *t* aspiratum, et ab hoc nomine *mathesis* media correpta, quod scientiam designat, derivari a multis refertur auctoribus, et certum est ex Graeco; quia *matheo* verbum idem est quod disco, et *mathetes* est discipulus, et *mathesis* disciplina. Unde *mathematica* est disciplinalis scientia et doctrinalis, sicut Cassiodorus dixit superius. Sed vocabulum falsae mathematicae sine aspiratione scribi asseritur ab eisdem auctoribus et a *mathesi* media producta, quod divinationem notat, descendere, vel quod certius est, a *mantos* vel a *mantia*, quae sunt idem quod divinatio, sicut Hieronymus dicit in originali ix Isaiae.

10 Rutkin, "Is Astrology a Type of Divination?" 36–74.

11 One should consult here William's insightful and foundational article, "Roger Bacon and the Edition of the Pseudo-Aristotelian *Secretum secretorum*," 57–73.

12 Bacon, *Opus maius*, vol. I, 239–240. Quicquid vero sit de hac scriptura et derivatione, tamen falsa mathematica est ars magica. Nam numerantur quinque species artis magicae, scilicet mantice, mathematica, maleficium, praestigium, sortilegium. Mathematica ergo est secunda pars artis magicae. Haec sibi usurpat considerationem coelestium characteribus, carminibus, coniurationibus, sacrificiis superstitionis, et fraudibus variis deformatam.

13 Bacon, *Opus maius*, vol. I, 240. Et ponit per virtutem constellationum omnia de necessitate contingere, nihil ad utrumlibet, nihil a casu nec fortuna, nihil a consilio.

14 I discuss these and related arguments more fully in Chapter 4 of my vol. I, and what we would call Bacon's magic with talismans and powerful words in Chapter 8.

15 I discuss this further in Chapter 6 of my vol. I.

16 I discuss this further in Chapter 8 of my vol. I, as does Weill-Parot in his foundational, *Les <<images astrologiques>> au Moyen Âge et à la Renaissance.*

17 I discuss the four types of astrological praxis in the excursus to the overall introduction of my vol. I.

18 Hackett, "The Attitude of Roger Bacon to the *Scientia* of Albertus Magnus," 53–72. I provide bibliography for the others in my vol. I, p. 42.

19 Bacon, *Opus tertium.*

20 Bacon, *Opus tertium*, 37. Et necesse est omnia sciri per hanc scientiam, quia omnes actiones rerum fiunt secundum specierum et virtutum multiplicationem ab agentibus huius mundi in materias patientes; et leges huiusmodi multiplicationum non sciuntur nisi a perspectiva, nec alibi sunt traditae adhuc; cum tamen non solum sint communes actioni in visum, sed in omnem sensum, et in totam mundi machinam, et in coelestibus et in inferioribus. Haec autem scientia non est adhuc lecta Parisius, nec apud Latinos, nisi bis Oxoniae in Anglia; et non sunt tres qui sciant eius potestatem: unde ille, qui fecit se auctorem, de quo superius dixi, nihil novit de huius scientiae potestate, sicut apparet in libris suis, quia nec fecit librum de hac scientia, et fecisset si scivisset, nec in libris aliis aliquid de hac scientia recitavit. For an insightful recent article on *perspectiva* in Bacon's works, see Mantovani, "The First of All Natural Sciences," 101–134.

21 Bacon, *Opus tertium*, 38. Et quae de perspectiva narravi modo patent manifeste ex Opere Majori, et tractatu quem collegi de perspectiva, qui est pars quinta principalis illius operis; et simul cum ea consulenda est magna pars quartae partis totius operis, scilicet ubi de multiplicatione specierum et virtutum agentium determinavi, quam licet ascribo ibi geometriae propter rationes suas, tamen extraxi de scientia perspectivae.

22 Bacon, *Opus tertium*, 38. [E]t hoc ideo, quia summa et principalis radix sapientiae; et pro philosophia et pro theologia, est in istis multiplicationibus, et infinita pulchritudo; et nec perspectiva, nec aliquid de philosophia sciri potest sine hoc. Sed ille, qui multiplicavit volumina, ignorat has radices: nam nihil de eis tangit; et ideo certum est ipsum ignorare res naturales, et omnia quae de philosophia sunt; et non solum ipse, sed totum vulgus philosophantium, quod errat per ipsum.

23 Bacon, *Opus maius*, vol. I, 300–301. Nam res mundi sciri non possunt nisi per notitiam locorum in quibus continentur. Locus enim est principium generationis rerum, ut dicit Porphyrius; quia secundum diversitatem locorum est diversitas rerum; et non solum naturalium, sed moralium et scientialium, ut videmus in hominibus quod secundum diversitatem regionum habent mores diversos et occupant se in artibus et scientiis diversis. Quia igitur philosophia intromittit se rebus mundi, multum ei deest adhuc apud Latinos, [postquam?] <*propter*?> non habet certificationem locorum mundi. Sed haec certificatio stat in cognitione longitudinis et latitudinis cujuslibet loci; tunc enim sciremus sub quibus stellis est quilibet locus, et quantum a via solis et planetarum, et quorum planetarum et signorum loca recipiant dominium, quae omnia faciunt diversas complexiones locorum: quae si scirentur, possit homo scire complexiones omnium rerum mundi et naturas et proprietates quas a virtute loci contrahunt.

24 Bacon, *Opus maius*, vol. I, 137–138. [C]omplexiones locorum mundi inveniri non possunt, nisi homo sciat leges multiplicationum praesignatas, quoniam nec falsum vitabit, nec verum poterit confirmare. Sed locus est principium generationis, quemadmodum et pater, ut dicit Porphyrius. Et nos videmus, quod omnia variantur secundum loca mundi diversa.

25 Bacon, *Opus maius*, vol. I, 138. Sed quoniam res huius mundi in diversis locis constitutae, quantumcunque propinquae sunt, recipiunt conos diversarum pyramidum venientium a toto coelo objecto eis, ideo accidit infinita diversitas. Nam ad singula puncta terrae veniunt coni pyramidum singularum, et quilibet punctus est centrum unius Horizontis novi.

26 Bacon, *Opus maius*, vol. I, 138. Et ideo videmus, quod duae herbae simul nascuntur de terra sine medio, et ideo duo gemelli in ventre matris sortiuntur diversitatem complexionis, ut postea mores habeant diversos, et sequantur artes diversas, et occupationes difformes per totam vitam. Et ideo virtutes coelorum et stellarum producunt ubique diversas res in proprietatibus et naturis, et in rebus generatis secundum propagationem [....] Et cum puer in nativitate exponitur aeri novo tanquam alteri mundo, tunc recipit conos pyramidum coelestium secundum singulas partes, et sic recipit impressiones novas, quas nunquam dimittit, quia quod nova testa capit, inveterata sapit. Et tunc confirmatur complexio radicalis quae semper manet usque ad finem vitae, licet complexio currens mutetur tota die. Et ad hanc radicalem complexionem sequuntur inclinationes ad mores et ad scientias et ad linguas, et ad quaecunque artificia et negotia, et ad omnem diversitatem quam videmus in omnibus.

27 Lindberg has written a great deal on Bacon's understanding of *perspectiva*, the bibliography for which will be found in my vol. I and in Mantovani's article cited above.

28 Bacon, *Opus maius*, vol. I, 119. Et haec est figura, quam specialiter elegit natura in omni multiplicatione et actione, et non quamcunque pyramidem, sed illam cujus basis est superficies agentis, et cujus conus cadit in aliquod punctum patientis, quia sic potest a tota superficie agentis species venire ad singula puncta patientis per singulas pyramides et infinitas.

29 Bacon, *Opus maius*, vol. I, 117. Et oportet quod multiplicatio fiat sphaerice. Nam agens multiplicat se aequabiliter in omnem partem, et secundum omnes diametros, et omnes differentias positionis, quae sunt sursum, deorsum, ante, retro, dextrorsum, sinistrorsum. Ergo undique exeunt lineae in omnem partem ab agente tanquam a centro.

30 Bacon, *Opus maius*, vol. I, 250. [A]d omne punctum terrae incidit conus unius pyramidis virtuosae a toto coelo. Et coni isti sunt diversae in natura, et pyramides similiter, quia diversas habent bases propter diversitates horizontum, quoniam quilibet punctus terrae est centrum proprii horizontis.

31 Albert the Great, *De natura locorum*. See Weisheipl, O.P., Appendix I: "Albert's Works on Natural Science (*libri naturales*) in Probable Chronological Order," 565–577, for the dating of his Aristotle commentaries. No. 4 on pp. 566–567 is on *De natura loci*. I also treat this material on Albert more fully in Chapter 2 of my vol. I.

32 De natura locorum, quae provenit ex habitudine loci ad caelum, tractaturi primo mentionem de his quae in PHYSICIS determinata sunt; in his enim probatum est, quod locus est generationis principium activum quemadmodum pater (1, 9–13).

33 Albert the Great, *De natura locorum*, 2. Ex omnibus ergo his satis liquet, quod oportet scire naturam loci, nec sufficit tractatus, qui in PHYSICIS habitus est de ipso, eo quod ille non nisi universaliter certificat de loco [....] Propter quod tradentes scientiam naturalem et non inducentes aliquid de locorum diversitate peccant et solacium quaerunt suae imperitiae, si dicunt non oportere de ipso quaerere; non enim sufficit, quod in MATHEMATICIS de huiusmodi diversitate traditum est, ubi traditur, quae stellae super quae loca hoc vel illo modo oriantur vel occidant, utrum videlicet oblique vel recte, et quae ipsarum nec oriantur nec occidant, sed semper appareant, vel quae numquam et quae oriantur et occidant aliquando; quia licet ex his sciatur situs climatum, non tamen per eadem sufficienter scitur natura locorum et diversitas complexionis ipsorum.

34 Albert the Great, *De natura locorum*, 8. Si quis autem particulariter velit cognoscere omnes naturas et proprietates particularium locorum in aqua et aere et terra, sciet, quod non est punctus in eis, qui non habeat specialem proprietatem ex virtute stellarum mediam habitationem commixtorum respicientium; ad quodlibet enim punctum habitationis animalium et plantarum et lapidum variatur circulus horizontis, et ad variationem circuli horizontis totus respectus caeli ad medium habitationis variatur. Qua de causa variantur naturae et proprietates et mores et actus et species eorum quae in eodem loco sensibili videnter generari, in tantum, quod et geminis seminibus et animalibus brutis et hominibus ex hoc diverso respectu proprietates et mores diversi attribuantur. Et hoc est rationabile, quia COMPERTUM EST caelum diffundere virtutes formativas in omne quod est. Maxime autem diffundit eas per radios emissos a luminibus stellarum, et ideo consequens est, quod quaelibet figura radiorum et angulus diversas in inferioribus causet virtutes.

35 Albert the Great, *De natura locorum*, 8, 19. Cum igitur ex mutatione horizontis necesse sit mutari totum circulum et ex mutatione circuli tota mutetur figuratio radiorum, cumque quodlibet punctum habitationis unum speciale centrum constituat horizontis, necessario consequitur quodlibet punctum habitationis habere virtutes speciales, quibus informatur id quod locatur in ipso. Et haec causa est, quare nulla rerum generatarum invenitur omnino et per omnia in proprietatibus similis alteri. Et hic locus proprie est, de quo dicit PORPHYRIUS, quod est unum de individuantibus [....] [C]onstellatio est causans virtutem qualitativam stellarum, quae infunditur inferioribus [....] Non tantum autem quilibet locus particularis est alterius proprietatis quam alius, sed etiam quilibet per se diversarum successive efficitur proprietatum, eo quod per ortum et occasum stellarum super ipsum et diversam habitudinem earum ad locum quemlibet necesse est variari proprietates ipsius. Et haec causa est, quare mutantur res generatae ex suis continentibus.

36 Albert the Great, *De natura locorum*, 24. Propter quod etiam sapientes in astris circulum uniuscuiusque loci diligenter primo considerant et deinde praecipiunt considerare virtutes locorum quasi stellas secundas et ex his duobus coniunctis prognosticare de natura rei generatae; stellae enim per aliud et non immediate influunt virtutes et utuntur ad influxum suarum virtutum duplici medio, quorum unum est radius et secundum locus continens. Et per haec duo veniunt ad materiam rei generatae, quam formant et terminant virtutes influxae. Variatio autem situs stellarum est causa variationis figurae, quam describunt radii super circulum horizontis. Et haec est causa variationis virtutum locorum et eorum quae generantur in locis, et ideo necesse est scire has variationes philosophum naturalem.

Bibliography

Primary Sources

Albert the Great. *Opera Omnia*. Edited by Paul Hossfeld. *De natura locorum*. Monasterii Westfalorum: Aschendorff, 1980.

Roger Bacon. *Opus majus*. Edited by John Henry Bridges. *3 vols*. Oxford and Edinburgh: Clarendon, 1897–1900.

Roger Bacon. *Opus tertium* Edited and Translated by Nikolaus Egel. Hamburg: Felix Meiner Verlag, 2020.

Roger Bacon. *Opus tertium*. Edited by John S. Brewer. *Fr. Rogeri Bacon Opera Quaedam Hactenus Inedita, vol. I: Containing the Opus Tertium, Opus Minus, and Compendium Philosophiae*. London: Longman, Green, Longman, and Roberts, 1859.

Roger Bacon. *Secretum secretorum*. Edited by Robert Steele. *Opera hactenus inedita Rogeri Baconi*, fasc 5. Oxford: Clarendon, 1920.

Secondary Sources

Hackett, Jeremiah. "The Attitude of Roger Bacon to the *Scientia* of Albertus Magnus." In *Albertus Magnus and the Sciences: Commemorative Essays*, edited by James A. Weisheipl, 53–72. Toronto: Pontifical Institute, 1980.

Mantovani, Mattia. "The First of All Natural Sciences. Roger Bacon on Perspectiva and Human Knowledge." *Vivarium* 59/3 (2021): 186–214.

Rutkin, H. Darrel. "Is Astrology a Type of Divination?: Thomas Aquinas, the Index of Prohibited Books and the Construction of a Legitimate Astrology in the Middle Ages and the Renaissance." *International Journal of Divination and Prognostication* 1 (2019): 36–74.

Rutkin, H. Darrel. *Sapientia Astrologica: Astrology, Magic and Natural Knowledge, ca, 1250–1800, I Medieval Structures (1250–1500)*. Springer, 2019.

Weill-Parot, Nicolas. *Les "images astrologiques" au Moyen Âge et à la Renaisance: Spéculations intellectuelles et practiques magiques (XIIe-XVe siècles)*. Paris: Honoré Champion Éditeur, 2002.

Weisheipl, James O.P. "Appendix I: Albert's Works on Natural Science (*libri naturales*) in Probable Chronological Order." In *Albertus Magnus and the Sciences: Commemorative Essays*, edited by James A. Weisheipl, 565–577. Toronto: Pontifical Institute, 1980.

Williams, Steven. "Roger Bacon and the Edition of the Pseudo-Aristotelian *Secretum Secretorum*." *Speculum* 69 (1994): 57–73.

7 Knowability and Power of *Ars*

Roger Bacon's Utopian Science as Anti-Magic

Nicolas Weill-Parot

The technological dreams of Roger Bacon are very well-known. They have earned the Franciscan friar comparisons with Leonardo da Vinci; he has often been seen as a kind of medieval Renaissance man. Bacon's discourse sounds as though he were pioneering modernity. But, as we know, a precise analysis seriously nuances this delusion, this fake feeling of being familiar with his views. His eschatological perspective has long been underscored, as has his anchorage in Franciscanism.[1] Furthermore, he is convinced that the majority of the extraordinary devices he praises were already achieved long ago in the past. Bacon gives a list of extraordinary inventions in several of his late works, the *Communia mathematica*,[2] *Opus maius*,[3] *Opus tertium*,[4] and especially the *Epistola de secretis operibus artis et naturae et de nullitate magiae*.[5]

I will not stress very well-known ideas about *scientia experimentalis*: his "experimental science" has three prerogatives: first, it certifies by experiment the conclusions of other sciences; second, it can reach "magnificent truths" located at the border of other sciences; and third, it uses the other sciences as handmaids in order to explain "secrets of nature."[6] I would like to reassess the issue concerning the opposition between, on the one hand, Bacon's *scientia experimentalis*, which is largely a utopian science, and, on the other hand, magic. I would like to grasp the deep roots of this opposition between this utopian science and magic.

My argument is that the keystone of this difference between magic and experimental science lies in the dialectic between occultation and knowledge – hence, the issue of "secrecy." There is a good secrecy in utopian science; there is a bad secrecy in magic. The distinction between good and bad secrecy lies in the issue of knowledge and truth. As we know, the theme of the "secret" is important in Bacon's late works.[7] I would like to stress this point, since I maintain that secrecy is more than important; it is a key to his scientific worldview. Secrecy is generally understood in Bacon's later works, notably in the *Opus maius*, as a provisional state of ignorance, rather than a definitive one. A secret is knowledge locked up in a chest. Just open the chest, and you will know that the secret will no longer exist. This is the main difference between secrecy and natural occult (i.e., occult properties).[8]

DOI: 10.4324/9781003405702-8

7.1 Three Frameworks

If we seek the role of secret in Bacon's idea of knowledge, there are three frameworks to keep in mind. First there is the Aristotelian framework. This implies that science consists of showing the natural causes as they are manifest to the senses. According to the *Posterior Analytics* I.2 (71b10-13):

> We think we understand something *simpliciter* (and not in the sophistical way, incidentally) when we think we know of the explanation because of which the objects holds that it is its explanation, and also that it is not possible for it to be otherwise.[9]

Science looks for explanations by manifest causes. There is no room for secrecy.[10]

Second, the scholastic framework. While presenting itself as Aristotelian, it includes a number of changes to its original model. The scholastic authors made room for several kinds of phenomena whose causes are definitively hidden. Galen and then Avicenna argued that certain kinds of natural bodies act neither through their primary qualities nor by the mixing of these qualities (namely their "complexion"), but through their whole substance or specific form, which gave them an occult property. For instance, a magnet attracts iron through an occult property. The occult property is not a secret which could be disclosed; it is something definitively unknowable. This model can be found in Albert the Great's closed rational worldview: everything natural can be known by reason, and the unknown is definitively enclosed in the concept of occult property.[11]

Third there is Bacon's framework. It is an open, rational framework where the unknown can be discovered by investigation and experiment. This is the positive value of the secret of art and nature. Bacon follows the ancient tradition of nature hiding its secrets. In the *Opus maius* and his *Epistola*, Bacon writes that we do not write "secrets of Nature on goats or sheep" to be understood by everyone.[12] The first meaning is obvious: Bacon is alluding to the parchments on which science is written. But, beyond that, the formulation is inevitably made to suggest a metaphor, an imaginary vision, a representation of an alternative world where natural truths would be written on the bodies of the goats, so that everyone could see them.[13] This nature that hides its secrets thus gives the implicit order to the *sapiens* philosopher not to reveal the greatest secrets to the crowd. Quoting Marbodus and his book of stones (*De lapidibus*), Bacon repeats that you must not "give pearls to pigs," to the crowd of ignorant people (including, according to Bacon, academic scholars following the traditional scholastic path). Quoting the *Secretum secretorum*, Bacon asserts that "he would break the heavenly seal, if he communicated the secrets of nature and art"; adding that many woes follow for the man who reveals secrets. Bacon says that the "ignorance of secrets arts and science" is one of the reasons for the bad things which affect humankind.[14]

Those secrets of nature and art are the pioneering territory of experimental science (*scientia experimentalis*). The *scientia experimentalis* aims at disclosing these secrets – its goal is knowledge. The follower of experimental science seeks to fully understand the whole natural process at work in a secret. His aim is to find the whole cause of what was secret. *Scientia experimentalis* is a driving force towards knowledge. The secrets are the provisional stages of this progression towards knowledge.

7.2 Magic

On the opposite side, there is magic. Magic also has to do with concealment. Like other medieval authors, Bacon certainly would agree that magic refers to astonishing artificial operations whose cause is hidden.[15] Concealing the cause is the keystone of magic. When the cause is elucidated, magic disappears. The first kind of hidden cause is a trick, and this is prestidigitation. Bacon criticises the *ioculatores* or prestidigitators, especially in the *Epistola de secretis artis and naturae*, because they deceive people by fake operations.[16] The second kind of hidden cause relies on demons, and such a magic is called *nigromancy* (the word "nigromanticus" occurs in the *Opus maius*[17]) – or angels (theurgy), according to some practitioners.

The third kind of hidden cause in magic, according to many medieval philosophers, can be a hidden natural force. This is the definition of natural magic, which articulates nature and art (*ars*) according to a natural hidden cause. But there is no room for natural magic in Bacon's views, that is, there is no room for a magic based on a hidden natural cause, because Bacon substitutes the *scientia experimentalis* for natural magic in order to account for the wonderful products of art and nature. We will come back to this important point.

The occultation in magic is very different from the secrecy of nature or art scrutinised by the *scientia experimentalis*. The magician's secrecy is bad secrecy, it is wrong secrecy, it is fake secrecy. It is a secret intended to deceive men. In the *Opus maius*, as in Bacon's other works, magic is always rejected. Some modern scholars misunderstand this point, but Bacon explicitly rejects magic everywhere. The issue lies in the fact that some practices that other authors would call magic are not considered to be magic by Bacon. This is not a rhetorical trick on Bacon's part – this distinction between magic and science is very consistent in all his writings. If we limit our investigation to the *Opus maius*, magic is constantly associated with fraud, superstition, and falsehood: Bacon speaks out against the *fraudulentia* of the magical art or the *falsitas magicae mathematicae*; he argues that the opinion of magicians on fascination is *vana*.[18]

The section devoted to fascination and incantations is significant.[19] Fascination, says Bacon, can be understood in two ways: according to the *vetulae sortilegae* and *magici* (old sorceresses and magicians), and according to a naturalistic explanation, since some men have bad complexions and then are contagious.[20] Bacon underlines the role of the influence of a certain constellation at a person's birth. He refers to his optical theory of multiplication of *species* in order to account for what looks like an action at a distance. He also insists on the power of *intentio* (intention), *desiderium* (desire), and *confidentia* (trust). To emphasise the naturalness of this process, he compares it to the scamonea that increases the effectiveness of a drug.[21] And he concludes, "*et adhuc non est magicum*" ("and again this is not magical"). Even the effectiveness of the utterance of words can have a naturalistic explanation, by referring, in particular, to the power of the constellations. But the *magici* and *vetulae maledictae* (magicians and old damned wives) do not pay attention to astrological time, or to natural features which explain the efficacy of words. They do not grasp the real cause of the phenomenon. Thus their opinion (*sententia*) is vain (*vana*).[22]

The essential reproach that Bacon makes against the magicians is that *they ignore the real causes*. The secrecy of magicians is no more than ignorance. Again, when he describes the wonderful properties of natural bodies (magnets, some plants, etc.), he adds that the magicians include some pointless incantations (*carmina*), and they wrongly think that these incantations are the cause of the wonders.[23] As we see again, magical secrecy is always a superfluous veil, useless, aiming only at deceiving men. Bacon's analysis of magic is not very demonological (there are some rare allusions to demons). More fundamentally, what Bacon reproaches magic for is its *falseness*. And the magic secret is precisely a veil of falsity. The magician himself misunderstands the true causes of things. The use of procedures that he considers secret (incantations, etc.) is a manifestation of his ignorance.

In the famous section of the *Opus maius* devoted to *scientia experimentalis*, Bacon argues that *scientia experimentalis* is the only way to "experiment perfectly what can be done by nature, what by the industry of art, what by fraud," and "what charms, conjurations, invocations, prayers, sacrifices, which are magic, mean and dream of," and "what is done in them in order to remove all that is false and to keep the sole truth of art." He adds that only *scientia experimentalis* "teaches how to investigate all the foolishness of the magical operations, not to confirm them, but to avoid them, just as logic investigates the art of sophisms."[24] The comparison with logic is obvious: logic is able to detect an error in a logical argument. Similarly, *scientia experimentalis* is able to discard the wrong practices and the wrong causes in a magical process. This does not mean that Bacon thinks that something can be saved in magic *qua* magic. This means that, according to Bacon, magicians can use some natural processes without knowing the real causes of these processes, namely the true natural causes. But, we must repeat that, according to Bacon, magic *as it is magic* is totally, completely condemned.[25] As he writes later, *scientia experimentalis* is the science whose function is to determine what can be made by nature and what can be made by art. Thus, this science knows how to remove all "magical illusions" and errors in "charms, invocations, conjurations, sacrifices, and worship."[26]

In the *Opus maius* and elsewhere, magic and science are in conflict according to three points. First conflict: magic has pretended to have the philosophical dignity of science, which is why at the beginning of the Christian era, Christian theologians were suspicious of philosophy and science.[27] Second conflict: magic pretends to reach achievements that only true experimental science allows us to reach. The list of Bacon's extraordinary devices – self-moving chariots, flying machines, etc. – shows achievements that magicians claim to reach with their pseudo-art. One can think, for example, of the use of mirrors. Experimental science uses mirrors to observe what is happening very far from the observer.[28] Catoptromancy uses mirrors for divination.[29] One could multiply examples of this type. Third conflict: some practices can be split between magical use and truly experimental use. For instance, the use of music. The natural use of music against diseases must be distinguished from the magical one. Moreover, in the *Communia mathematica*, Bacon explains that in addition to the ordinary use of instruments, an experimental musician can use them in order to impose his will on other men – hence, a utopian use.[30]

Therefore, the competition between magic and experimental science, as we can see, takes place on several levels. We can briefly see the example of what was called "astrological images," namely, talismans supposedly endowed by the sole natural influence of stars and constellations. This notion was coined by the middle of the thirteenth century by the anonymous author of the *Speculum astronomiae*, and it gave rise to a long debate among theologians, philosophers, physicians, and astrologers.[31] In the *Opus maius*, Bacon quotes the "classical" sources for this art of images: the history of Moses who made an astrological ring to make his Ethiopian woman forget that she was married, the ninth *verbum* of pseudo-Ptolemy's *Centiloquium* with the commentary by Hali, which tells the story of a man healed from a scorpion sting thanks to the astrological seal of Scorpio, and the famous *De imaginibus* by Thebit.[32] But the crowd of students believe that these practices are beyond human intellect, whereas the philosopher knows that these works of wisdom (*opera sapientiae*) can be done through adequate constellations, which means that these "images" are based on a true knowledge of natural causes. Thus, almost no one dares to speak in public about these operations, because those who dare to do so are called magicians, whereas they are very wise and learnt people. It is because magicians abuse this knowledge that theologians and decreetists have rejected these practices by classifying them as magical practices.[33] But they are *not* magic.

The Magister Speculi, i.e., the author of the *Speculum astronomiae*, classified the *scientia imaginum* under the part of astrology dedicated to elections (the astrological determination of the best time for a specific action).[34] In the *Opus tertium*, in the part dealing with *scientia experimentalis*,[35] Roger Bacon explains how this master science dominates all the others and, in particular, how it makes use of astrology to produce special beverages, foods, or medicines according to specific constellations; these products are able to modify the complexion of men so that they will be inclined to act according to the will of the producer.[36] And "this," writes Bacon, "is the origin of all the philosophical images, charms, and characters." And he adds: "And this is the reason why this science," i.e., experimental science, "makes the distinction between those kinds" of objects. It reveals that some of them are "made according to the philosophical truth," whereas "others" are made "according to the fault and error of the magical art." Experimental science identifies "the types of magical art" and "separates them from the philosophical truth."[37]

Bacon's "philosophical images" probably refer to the same kind of practices as Magister Speculi's "astrological images" or Albert the Great's "seals," that is, magical objects deriving their power from the natural influence of the stars. Nevertheless, there are two great differences between Albert the Great's scholastic framework and Bacon's framework concerning those astral talismans. In his *De mineralibus*, Albert the Great clearly writes that this part dealing with astrological seals belongs to nigromancy, namely the part of nigromancy which is "subordinate to astrology."[38] This is to say, he asserts clearly that the art of astrological seals falls under magic, precisely astral magic, and more precisely natural astral magic. On the other hand, Bacon completely excludes the good part of "images," i.e., the "philosophical images," from the category of magic.

Second, Bacon does not exclude that "charms" and "characters" can also be "philosophical," whereas both the Magister Speculi and Albert the Great would maintain that charms and characters, i.e., linguistic processes, are "addressative," i.e., addressed to superior intelligences, i.e., demons. Therefore, only "images," which involve no charms or inscriptions of characters, can be called purely "astrological," and, hence, can be seen as licit.[39]

In *De mineralibus*, Albert the Great gives an explanation of the way that a seal carved on a gem can naturally acquire powers from the stars. The craftsman makes the astrological seal at a special astrological time, and the constellation endows the seal with a special power.[40] This special power corresponds to what was often called an occult property: just like some species (for instance, a magnet) receive at birth a property corresponding to its species or specific form, a man-made object can therefore receive such a property. The difference is that for talismans, this property is not specific or substantial, but individual or accidental. The occult property is the keystone for natural magic, since it is something definitively unexplained but which receives a concept.[41]

Finally, Albert the Great builds up a naturalistic explanation of astrological seals or talismans based on three principles: (1) *non-addressativity*: that means that in their making, there are no rituals, no inscriptions, or other signs addressed to demons; (2) the idea, shared by most Medieval thinkers, that human *ars* (human artificial making) cannot equal nature; (3) the idea that the source of the power acquired by the astrological seal is a property (which later will be called an occult property).

Bacon's framework is different:

- The issue of addressativity is irrelevant in Bacon's view since Bacon can assert that not only can powerful images receive a naturalistic explanation, but so can words and charms.
- Bacon is one of the very few medieval authors who asserts that *ars* can surpass nature.[42]
- Bacon's views of nature are not based on occult (definitively unknown), but rather on secret (temporarily unknown).[43]

In such a framework, there is no room for natural magic but for an experimental science whose principle is *knowability*. Human *ars* is powerful enough to endow man with the complete mastery of a mechanism to surpass nature. And thus, human intellect will be able to understand the complete causative process. This causative process will appear in full light.

7.3 Experiment

The "philosophical" images, just like the philosophical charms and characters, are not magic because they are based on true knowledge, and this knowledge is primarily based on experiment. We know that Bacon greatly praises Peter Peregrinus (Pierre Pèlerin de Maricourt) as the best *experimentator* of his time. This prominent figure of medieval science, a non-academic scientist sometimes described as a "military engineer," wrote the famous *Epistola de magnete* in 1269. Historians of science always make mention of his name because he is one of the few medieval scientists who actually performed real experiments according to the standards of modern science. Indeed, in his *Epistola*, he describes several experiments made in order to understand magnetic polarity. But what is interesting is that, insensibly, without any solution of continuity, he comes to describe imaginary inventions. After describing true experiments with parts of a magnet in order to discover the rules of polarity, he describes a self-moving sphere made of lodestone; and after describing different kinds of compasses, he suggests a perpetual wheel moved by a magnetic device. These two imaginary inventions are put forward, in keeping with the true experiments with magnets, without any discontinuity.

This process of continuous extrapolation outside of what is actually or currently possible, and which presents itself as remaining within the framework of science, corresponds precisely to what I have called *scientific utopia*. It is more than a vague dream of extraordinary inventions. It is truly a projection of science beyond its limits without a complete acknowledgement that this projection transgresses the scientific borders.[44] From this standpoint, Peter Peregrinus gives indications that he considers these inventions to be on the border of the possible and the impossible, but without ever expressly acknowledging it. When he describes the self-moving sphere, he warns that if the device does not work, the failure must be attributed to the inability of its manufacturer rather than to nature.[45] This is a way of implicitly or unconsciously admitting the oscillation between the possible and the impossible, which is a distinctive feature of this "scientific utopia."

Now, let us briefly examine this self-moving magnetic sphere. Peter Peregrinus rejects the wrong opinion of those who ascribe the magnetic force to the mineral places where there are lodestones or to the North Star. He attributes the origin of this force instead to the celestial poles; this explanation is the most significant improvement in the understanding of magnetism until 1600 when William Gilbert demonstrated that the origin of the magnetic force is located in the *terrestrial* poles.

But Peter adds that it can be postulated that all the parts of the lodestone receive their power from the other parts in the heavens. Therefore, we should make a magnetic ball and fix it between a pivot up and a pivot down, that is, on the two poles of the round lodestone. The whole weight must be equally distributed. The magnetic sphere must be located on the axis of the meridian, each pivot being at the corresponding poles. "If you do that," argues Peter, the lodestone "will turn according to the motion of the heavens." So you will "be delighted that you have discovered a wonderful secret."[46] The word "secret" is important, as it has almost the same meaning as Bacon's secret of art and nature.

Peter ascribes an astronomical goal to this turning magnetic sphere: it will serve as an armillary sphere or astronomical clock, which will naturally show the exact astrological or astronomical chart. Therefore, you will no longer need a timepiece. Thus, Peter assumes that the craftsman has already carved a complete celestial chart on the sphere, showing planets, stars, and constellations. He is describing a kind of armillary sphere or astronomical clock which moves itself according to the motion of the celestial sphere.

Now, as Robert Halleux has pointed out, Roger Bacon mentions this self-moving sphere in the *Opus maius*.[47] These lines occur in the section devoted to the second prerogative of *scientia experimentalis*, namely its capacity to find great truths located in the limits of other sciences. The first example tells that mathematics can produce an instrument able to show longitudes and latitudes; but it is not in the power of mathematics to make this instrument move naturally according to the diurnal movement. "A perfect experimenter can study the ways for such a motion." He can be inspired by the natural bodies which turn or move according to the celestial motions: the three elements (except earth), comets, seas, rivers, medulla, brain, the matter of diseases, and the plants. Bacon asserts that such an instrument would be as valuable as a king's treasure, and it would replace all astronomical instruments. But few people want to seek such "miracles" at the borders of mathematics. This self-moving sphere can be found in other works by Bacon, notably the *Communia mathematica* and *Epistola de secretis artis et naturae*.[48]

Thus, the field of experiments, or better the *scientia experimentalis*, allows a continuity between truly experimental phenomena (lodestone, the fact that plants open or close their parts according to the motion of the Sun, etc.) and utopian devices (such as the self-moving astronomical sphere). This process can be called extrapolation. In the list of extraordinary instruments in the *Epistola*, Bacon mentions "an instrument three fingers high and just as wide, or of less size," by which "a man could snatch up himself and his friends and raise and lower them from every danger of prison."[49] Here, Bacon extrapolates from the works of statics by the great thirteenth-century scientist and mathematician Jordan of Nemore.

Even when the description of the extraordinary machine is only sketched out, as, for example, the machine for flying mentioned in the *Epistola*, it is predicated that when this machine is produced, all its causes will be known and elucidated.[50] In other words, the secret of art and nature concerning it will have been completely disclosed.

7.4 Three Ways of Occulation and Bacon's Understanding of the Philosopher's Task

To grasp the peculiar conception of concealment in Bacon's views, three different types of concealment are to be differentiated.

The first is that of the natural occult. It is an unsurpassable unknown, contained once and for all by the concept of an occult property. This is the dominant form of occultation in scholastic philosophy. Such a concept can be found in Albert the Great, Thomas Aquinas, and later in many medical writings by Arnald of Villanova, Peter d'Abano, etc. To be sure, Bacon does not deny the existence of occult property, but the framework he puts forward is not based on this notion, as it is, for instance, in Albert the Great. In Albert the Great's views, in particular, this notion corresponds to a natural world already wholly explainable by Aristotelian philosophy, since the only unexplainable phenomena are definitively contained and encapsulated within these notions of occult property and specific form. This conveys a worldview of a rational and closed science, with the idea that natural philosophy is already able to explain the whole natural world.[51]

The second type of concealment is that of the magicians. The magicians conceal by false secrets and vain, superfluous, useless procedures which aim only at deceiving the crowd, and which deceive even the magicians themselves, since they are misled about the true causes of things.

The third mode of occultation is that of Bacon's secrets of art and nature. These secrets are those that experimental science aims to elucidate. They are thus provisional secrets, destined to be revealed by the experimentator and the philosopher; in short, by the one who implements the resources of *scientia experimentalis*. In contrast to the complete and closed scientific rationality of Albert the Great, Bacon's scientific rationality is open: science is unfinished, and it is the undefined territory of the secrets of nature and art (*secreta naturae et artis*) that is the realm of its undefined future. Hence, a philosopher may be led to hide a natural truth from the crowd of non-knowers by a secret; but this secret will then be legitimate, since he will be obeying the divine desire of a nature that hides its secrets from those who are not worthy of them.[52]

Now, occultation is essential to the definition of magic, whether it is natural, as in Albert the Great's theory of astrological seals, or nigromantic, as in the case of magicians. If the cause is no longer hidden, there is no more magic. On the contrary, experimental science rests entirely on the postulate of a complete knowledge of the causal chain, allowing the achievement of this or that extraordinary machine.

This is why I put forward the concept of "scientific utopia" to give a name to this impulse pushing science to exceed its own limits beyond what is currently possible. Some researchers have taken issue with the supposed anachronicity of this term. While one must certainly refrain from imposing external models on medieval texts, this is a concept that I have drawn from the study of texts: Roger Bacon, of course, but also Peter Peregrinus, the writings describing perpetual wheels, such as that of Jacques Legrand, etc.[53] And this concept seemed absolutely relevant. Wanting to stick to the only lexicon used can sometimes lead to going in circles. A notion can haunt a reasoning without being named.[54]

As stated above, in generic terms, a "scientific utopia" can be defined as follows: (1) it is an *imaginary* projection involving a technological dream or thought experiment; (2) it is astonishing and *extraordinary*, inducing wonderment or fear; (3) it implies an attempt at *rationalization* – at least a slight one. As the term suggests, a scientific utopia is part of scientific thought. This is precisely the way used by Bacon. It comes up against a scientific or technological limit which makes the project unrealistic, maybe for quite a long time; (4) a scientific utopia is *possible and yet impossible*: its actual potential for realisation is indeterminate. Bacon locates his extraordinary instruments in an uncertain place between fiction, the past, present, and future. This possibility remains indeterminate, undefined; (5) the utopian scientist is implicitly or explicitly *aware* that his idea is impractical at the time when he lives. Bacon assures of the present possibility of realising the extraordinary instrument, but this assertion itself reflects this awareness of a distance from the stage of their realisation.[55]

In summary, a scientific utopia involves an imaginary and extraordinary project, expressed within the rational framework of science; its status wavers between the possible and the impossible in the thinker's own view. It is an intentional projection beyond what is currently viewed as possible for science. I cannot insist on the specific scientific utopia in Bacon's eschatological project of reformation of Christianity (a question I have dealt with elsewhere). I will only remind that Bacon makes extrapolations from the science of his time (for instance, statics) and that he believes that most of these extraordinary inventions were discovered in ancient times and have been forgotten. In a way, in Bacon's mind, most of the scientific utopias are retro-utopias.[56] The idea that art can surpass nature is one of the essential pillars of this utopian conception. This is also one of the major differences between Albert the

Great and Roger Bacon. Albert, for whom art can never even equal nature, leads to the notion of natural magic, where the natural cause remains definitively inexplicable. Roger Bacon's belief in the superiority of art's power over nature leads to the notion of experimental science, or more broadly, scientific utopia, where the cause is supposed to be clearly elucidated.

Thanks to the concept of scientific utopia, we can better understand the processes and the stake of Bacon's "experimental science," particularly in its opposition to magic. Knowability is the line between the utopian experimental science and deceptive magic. That is why the concept of secret of art and nature is so important in the *Opus maius* and other late works by Bacon.

Notes

1 Easton, *Roger Bacon*; Beaujouan, "La prise de conscience de l'aptitude à innover," 5–14; Matus, "Reconsidering Roger Bacon's Apocalypticism," 189–222; Sannino, "Riforma degli studi e della società nel *Compendium studii Philosophiae* di Ruggero Bacone," 25–47; and Power, *Roger Bacon and the Defence of Christendom.*

2 Bacon, *Communia mathematica,* Part I, dist. 3, Chap. 2–3, 42–47.

3 Bacon, *Opus maius*, vol. II, Part 6, 217–218.

4 Bacon, *Opus tertium*, 18, ed. Little.

5 Bacon, *Epistola de secretis operibus artis et naturae*, Ch. 4, 532–533. Some scholars still have some doubts about the authenticity of this work; it seems that there are no strong reasons for it (except the absence of contemporary manuscripts and a chronologically impossible dedication to William of Auvergne), since, first, almost all the sections of this text can be found elsewhere in other authentic works by Bacon; second, the text is perfectly built as a consistent whole. In my opinion, it seems unlikely that a sectator of Bacon could build such a consistent work from copied pieces that perfectly fit with what we know, after many readings and studies, about Bacon's views. William Newman underlined that the doubt focused above all on the last chapters concerning alchemy, but he maintained that they may be authentic or, at least, may contain some authentic sections. Newman, "An overview of Roger Bacon's Alchemy," 328.

6 Bacon, *Opus maius*, vol. II, Part 6, 172, 202, and 215; Bacon, *Opus tertium*, Ch. 13, 43–45, ed. Brewer. See also Hackett, "*Scientia experimentalis*: From Robert Grosseteste to Roger Bacon," 89–120; Hackett, "Roger Bacon on Experimental Science," 277–315; Hackett, "Ego Expertus Sum," 145–173, and the chapter by Jeremiah Hackett in this volume.

7 Eamon, *Science and the Secret of Nature.*

8 Weill-Parot, "Encadrement ou dévoilement," 151–170; Weill-Parot, "Silence de l'auteur ou de la nature?" 83–91; and Weill-Parot, "*Secretus* et *occultus*," 1–20. For more on specific types of secrecy, see the chapters by Meagan Allen and William Newman in this volume.

9 Aristotle, *Posterior Analytics*, II.1, 71b10-13, 2. trans. Jonathan Barnes.

10 On the distinction between Greek classical science and Hellenistic science concerning secrecy, see Hadot, *Le Voile d'Isis*.

11 Weill-Parot, "Astrology, Astral Influences, and Occult Properties," 201–230; Weill-Parot, *Points aveugles de la nature.*

12 Bacon, *Opus maius*, vol. I, Part 1, Chap. 4, 10. Quoniam secundum sententiam Aristotelis libro secretorum et Socratis magistri sui, secreta scientiarum non scribimus in pellibus caprarum et ovium, ita quod a vulgo valeant aperiri. See also Bacon, *Epistola de secretis operibus artis et naturae*, 543. Sed considero quod in pellibus caprarum et ovium non traduntur secreta naturae.

13 This second interpretation is also obvious, despite the scornful criticism I received from an anonymous expert in an scientific journal .

14 Bacon, *Opus maius*, vol. I, Part 1, Chap. 4, 10. Nec oportet margaritas spargi inter porcos; nam rerum majestatem minuit, qui vulgat mystica: nec manent secreta, quorum turba est conscia, sicut in libro Gemmarum edocetur. Atque Aristoteles in libro secretorum dicit,se fore fractorem sigilli coelestis, si secreta naturae vulgaret.

15 This is the general definition I put forward because it embraces almost all the known concepts of magic in the Middle Ages (we can except William of Auvergne and a few others, who call magic even astonishing natural phenomena in which there is no human intervention, but generally, these kinds of phenomena are seen as natural wonders in scholastic thought). On William of Auvergne and natural magic, see Sannino, *Reading William of Auvergne*, 76–81.

16 Bacon, *Epistola de secretis artis et naturae*, Ch. 1, 523–524.

17 See, e.g., Roger Bacon, *Opus maius*, Part 7, 397.

18 Bacon, *Opus maius*, vol. I, Part 1, Chap. 14, 30. evacuata artis magicae fraudulentia; ibid, Part 4, 248: evacuata falsitas magicae; ibid Part 4, 399 (see note 22).

19 Bacon, *Opus maius*, vol. I, Part 4, 398. On Fascination, see Delaurenti, "La fascination et l'action à distance," 137–154; and Delaurenti, *Fascination*.

20 Bacon, *Opus maius*, vol. I, Part 4, 398. Et iterum potest elici quid intelligendum est de fascinatione, quae licet habeat verbum suspectum, et potest modo intelligi ut vetulae sortilegae et magici considerant: tamen certum est quod multi homines sunt malae complexionis et corruptae compositionis et infirmae, ita quod contagiosi sint, sicut homines infirmi multis morbis contagiosis. Et praecipue accidit complexioni hujusmodi virtus infectiva et contagiosa per malam constellationem in qua conceptus est puer vel natus.

21 Bacon, *Opus maius*, vol. I, Part 4, 398. Nam dicunt sapientes, quod sicut scammonea acuit medicinam et dat ei vigorem, sic intentio et desiderium et confidentia se habent in operibus humanis, et adhuc non est magicum.

22 Bacon, *Opus maius*, vol. I, Part 4, 399. Sed quia magici et vetulae maledictae non considerant fascinationem per has vias, immo asserunt alterationes subitas indifferenter fieri per homines quoscunque, cum videtur res in mane vel sero, vel loquuntur de illa, ideo vana est eorum sententia et a sapientibus reprobata.

23 Bacon, *Epistola de secretis artis et naturae*, Ch. 2, 525. Sicut si omnino esset ignotum quod magnes traheret ferrum, et aliquis volens hoc opus perficere coram populo, faceret characteres et carmina proferret, ne perciperetur quod totum opus attractionis esset naturale, hoc totum opus esset error.

24 Bacon, *Opus maius*, vol. II, Part 6, Chap. 2, 172. Haec [scientia experimentalis] ergo sola novit perfecte experiri quid potest fieri per naturam, quid per artis industriam, quid per fraudem, quid volunt et somniant carmina conjurationes invocationes deprecationes sacrificia, quae sunt magica, et quid fit in illis ut tollatur omnis falsitas et sola veritas artis teneatur. For this paragraph, I have taken into account the English translation by Robert Belle Burke, 587.

25 For instance, it seems that Amanda Power (*Roger Bacon and the Defence of Christendom*) misses this point, e.g., p. 49: "He developed a strong faith in the power of disciplines such as astrology, alchemy and magic". These three disciplines are not at all on the same level for Bacon: while he greatly approves of astrology and alchemy, he totally rejects magic *as magic*. I do not agree, either, with Steven Marrone's presentation of Bacon's attitude towards magic in his otherwise brilliant

and stimulating book *A History of Science, Magic and Belief from Medieval to Early Modern Europe*, 139–140. Indeed, Bacon does not accept "much of scientific magic": if we read carefully what Bacon writes, it is clear that he rejects all magic as magic, and the parts of magic he saves are understood by him *not as magic* but as an effect of such true sciences as experimental science or astrology.

26 Bacon, *Opus maius*, vol. II, Part 6, 221. Est autem alius modus utilissimus; quoniam ad hanc, ut dixi, scientiam pertinet judicare quid potest fieri per naturam aut per artis industriam, et quid non. Et novit illa separare magicas illusiones, et deprehendere omnes earum errores in carminibus et invocationibus et conjurationibus et in sacrificiis et culturis.

27 See note 33.

28 Bacon, *Epistola de secretis operibus artis et naturae*, Ch. 5, 534.

29 On catoptromancy, see Boudet, *Entre science et nigromance*, 101–106.

30 Weill-Parot, "Des instruments de musique concurrents de la magie: Roger Bacon et le pouvoir de la musique," 221–235.

31 Weill-Parot, *Les « Images astrologiques » au Moyen Âge et à la Renaissance.*

32 Bacon, *Opus maius*, vol. I, Part 4, 394.

33 Bacon, *Opus maius*, vol. I, Part 4, 394. Sed quia haec opera videntur vulgo studentium esse supra humanum intellectum, quia vulgus cum suis doctoribus non vacat operibus sapientiae, ideo vix est aliquis ausus loqui de his operibus in publico. Statim enim vocantur magici, cum tamen sint sapientissimi qui haec sciunt. Utique theologi et decretistae non instructi in talibus, simulque videntes quod mala possunt fieri sicut et bona, negligunt haec et abhorrent, et computant inter magica. Vident enim quod magici et abusores documentorum sapientiae his utuntur, et ideo aestimant indigna Christianis. Sed non debet veritas damnari pro ignorantia, nec utilitas pro malitia, quamvis eadem fieri valeant; nam tunc deberent homines carere cultellis in mensa, quoniam comedentes possunt interficere socios eodem ferro quo scindunt cibaria.

34 *Speculum astronomiae*, Ch. 11, 240. Parti autem electionum dixi supponi imaginum scientiam, non quarumcumque tamen sed astronomicarum.

35 Little, on p. 47, wrongly added to the section this title "[*De scientia quinte essentie*]," but is obvious that the whole section is about *scientia experimentalis*.

36 Bacon, *Opus tertium*, 52, ed. Little.

37 Bacon, *Opus tertium*, 53, ed. Little. Et hic est origo omnium philosophicarum* ymaginum, et carminum, et caracterum. Et ideo hec scientia distinguit inter hujusmodi, reperiens aliqua secundum philosophie veritatem facta, et alia secundum abusum et errorem artis magice. Et revolvit species artis magice [magice *T**; magices ed.], et separat eas a veritate philosophie." *The variant "magice" is given by Little in a footnote according to ms. *T*: Oxford, Bodleian Library, Tanner 116. Winchester College, Fellow's Library ms. 39, f. 190v gives the abbreviation of "philosophicarum" as Little put it (consulted on the digitised copy of a microfilm at the IRHT, Paris.). But, on the other hand, in the *Epistola de secretis artis et naturae*, Ch. 5, 534, we read in Brewer's edition: "Sed magis philosophicae figurationes inveniuntur," but in a footnote the variant "Sed profecto mirandae satis inveniuntur figurationes physicae"" which in this context makes sense (as I will show in a forthcoming essay). Nevertheless, in this section of the *Opus tertium*, we can guess that it is really "philosophical images," since true philosophy is here opposed to magic.

38 Albert the Great, *De mineralibus*, II.3.1, Ms Berlin, Staatsbibliothek zu Berlin, Preußischer Kulturbesitz, Latin qu. 906, f. 46r-114v, here f. 80r: "De ymaginibus autem lapidum et sigillis post hec est dicendum: licet enim pars illa [ista Ed.] scientie pars sit nigromancie [necromantiae Ed.] secundum illam speciem nigromancie que astronomie subalternatur que etiam de ymaginibus et sigillis vocatur, tamen

propter bonitatem doctrine et, quia illud cupiunt a nobis scire nostri socii aliquid de hoc, hic dicemus, omnino imperfecta et falsa reputantes quidquid de his scriptum a multis antiquis [*om.* Ed.] invenitur" (between brackets, the significant variants in the edition Magni, *Opera omnia*, 48 col. 1). For more on Albert the Great and Roger Bacon, see the chapter by H. Darrel Rutkin in this volume.

39 Weill-Parot, "Astral Magic and Intellectual Changes," 167–187.

40 Weill-Parot, "Albertus Magnus, *De mineralibus*, III.3.3" (this work provides an edition based on the MS of Berlin and the edition of Borgnet); Weill-Parot, "Causalité astrale et "science des images," 207–240; and Weill-Parot, "The Magical Powers of Carved Stones," 255–266.

41 Weill-Parot, "Astrology, Astral Influences, and Occult Properties."

42 See, for instance, Bacon, *Epistola de secretis operibus artis et naturae*, Ch. 5, 523. Nam licet natura [naturae *ed.*] potens sit et mirabilis, tamen ars utens natura pro instrumento potentior est virtute naturali, sicut videmus in multis; and Bacon, *Opus maius*, vol. II, Part 5, 164. Et quod natura potest facere, ars perficiens naturam multo magis potest illud operari. See also Allard, "Réactions de trois penseurs du XIIIe siècle vis-à-vis de l'alchimie," 97–106.

43 See note 8.

44 Weill-Parot, "Roger Bacon et l'utopie scientifique au Moyen Âge," 45–96.

45 Petrus Peregrinus, *Epistola de magnete*, I.10, 220. Et si tunc lapis moveatur secundum celi motum, gaudeas te esse assecutum secretum mirabile; si vero non, imperitie tue, potius quam nature, defectus imputetur (another edition: Petrus Peregrinus, *Epistula de magnete*).

46 See note 45.

47 Bacon, *Opus maius*, vol. II, Part 6, 203. Sed quod hoc corpus sic factum moveatur naturaliter motu diurno, non est in potestate mathematicae. Experimentator autem perfectus potest considerare vias hujus motus, excitatus ad eas considerandas per multas res quae sequuntur motum coelestium, ut sunt primo tria elementa, quae rotantur circulariter per influentiam coelestem [....] Et tunc thesaurum unius regis valeret hoc instrumentum et cessarent instrumenta astronomiae, et horologia, et esset pulcherrimum spectaculum sapientiae. Sed pauci de tanto miraculo et similibus in mathematicae terminis praeclare et utiliter scirent cogitare. See also Hallеux, "Entre philosophie naturelle et savoir d'ingénieur," 3–17.

48 Bacon, *Communia mathematica*, I.3.3, 45; Bacon, *Epistola de secretis operibus artis et naturae*, Ch. 6, 537.

49 Bacon, *Epistola de secretis operibus artis et naturae*, Ch. 4, 533. Nam per instrumentum altitudinis trium digitorum, et latitudinis eorumdem, et minoris quantitatis, posset homo seipsum et socios suos ab omni periculo carceris eripere, et elevare, et descendere.

50 Weill-Parot, *Le Vol dans les airs au Moyen Âge.*

51 On this point, see my previous studies (note 8).

52 See the articles by Weill-Parot mentioned in note 8.

53 On perpetual wheels, see Lohrmann, "Idee und Wirklichkeit des Perpetuum mobile im Mittelalter," 227–251.

54 I wonder why my concept of "scientific utopia" (which was well-founded on a careful reading of texts) was rejected by the blind reviewers of the journals *Speculum* and *Isis*. I think this reflects the exclusion of the "conceptual history of science" so dear to Edward Grant. But research in history of science cannot be limited to blind erudition on the one hand (for the best), or to social history on the other (for the worst).

55 Weill-Parot, "Roger Bacon et l'utopie scientifique," 221–235

56 See note 55.

Bibliography

Primary Sources

Albert the Great. *De mineralibus*. In MS. Berlin, Staatsbibliothek zu Berlin, Preussicher Kultubesitz, Latin qu. 906, f. 46r–114v.

Albert the Great. *Opera omnia*. Edited by Auguste Borgnet. Paris: Ludovicus Vivès, 1905.

Aristotle. *Posterior Analytics*. Translated by Jonathan Barnes, 2nd edition. Oxford: Clarendon, 1993, reprint 2002.

Peter Peregrinus of Maricourt. *Epistola de magnete*. Edited by Patricia Radelet de Grave and David Speiser. "Le *De magnete* de Pierre de Maricourt. Traduction et commentaire." *Revue d'histoire des sciences* 28/3 (1975): 193–234.

Peter Peregrinus of Maricourt. "Epistolae de Magnete/ Nova Compositio Astrolabii Particularis." In *Petrus Peregrinus de Maricourt, Opera*, edited by Loris Sturlese and Ron B. Thomson. Pisa: Scuola normale superiore, 1995.

Roger Bacon. *Communia mathematica*. Edited by Robert Steele. *Opera hactenus inedita Rogeri Baconi*, fasc 16. Oxford: Clarendon, 1905–1940.

Roger Bacon. *Epistola de secretis operibus artis et naturae et de nullitate magiae*. In *Fr. Rogeri Bacon Opera Quaedam Hactenus Inedita, vol. I: Containing the Opus Tertium, Opus Minus, and Compendium Philosophiae*, edited by John S. Brewer. London: Longman, Green, Longman, and Roberts, 1859.

Roger Bacon. *Opus majus*. Edited by John Henry Bridges. *3 vols.* Oxford and Edinburgh: Clarendon, 1897–1900.

Roger Bacon. *Opus tertium*. Edited by John S. Brewer. *Fr. Rogeri Bacon Opera Quaedam Hactenus Inedita, vol. I: Containing the Opus Tertium, Opus Minus, and Compendium Philosophiae*. London: Longman, Green, Longman, and Roberts, 1859.

Roger Bacon. *Opus tertium*. Edited by Andrew G. Little, *A Part of the Opus Tertium*. Aberdeen: The University Press, 1912.

Secondary Sources

Allard, Guy H. "Réactions de trois penseurs du XIIIe siècle vis-à-vis de l'alchimie." In *La Science de la nature: théories et pratiques*, 97–106. Montréal and Paris: Bellarmin-Vrin, 1974.

Beaujouan, Guy. "La prise de conscience de l'aptitude à innover (le tournant du milieu du 13e siècle)." In *Le Moyen Age et la Science: approche de quelques disciplines et personnalités médiévales*, edited by Bernard Ribémont, 5–14. Paris: Klincksieck, 1991.

Boudet, Jean-Patrice. *Entre science et nigromance: astrologie, divination et magie dans l'Occident médiéval:XIIe-XVe siècle*. Paris: Publications de la Sorbonne, 2006.

Delaurenti, Béatrice. *Fascination. Une histoire intellectuelle du mauvais œil (1140–1440)*. Paris: éditions du Cerf, 2023.

Delaurenti, Béatrice. "La fascination et l'action à distance: questions médiévales (1230–1370)." *Médiévales* 50 (2006): 137–154.

Eamon, William. *Science and the Secret of Nature: Books of Secrets in Medieval and Early Modern Culture*. Princeton: Princeton University Press, 1994.

Easton, Stewart C. *Roger Bacon and His Search for a Universal Science*. New York: Columbia University Press, 1952.

Hackett, Jeremiah. "Ego Expertus Sum: Roger Bacon's Science and the Origins of Empiricism." In *Expertus sum. L'expérience par les sens dans la philosophie naturelle médiévale*, edited by Thomas Bénatouil and Isabelle Draelants, 145–173. Florence: SISMEL-Edizioni del Galluzzo, 2011.

Hackett, Jeremiah. "Roger Bacon on Experimental Science." In *Roger Bacon and the Sciences – Commemorative Essays,* edited by Jeremiah Hackett, 277–315. Leiden: Brill, 1997.

Hackett, Jeremiah. "*Scientia Experimentalis*: From Robert Grosseteste to Roger Bacon." In *Robert Grosseteste: New Perspectives on his Thought and Scholarship*, edited by James McEvoy, 89–120. Turnhout: Brepols, 1995.

Hadot, Pierre. *Le Voile d'Isis: Essai sur l'histoire de l'idée de nature.* Paris: Gallimard, 2004.

Halleux, Robert. "Entre philosophie naturelle et savoir d'ingénieur: l'*Epistola de magnete* de Pierre de Maricourt." *Archives internationales d'histoire des sciences* 56/156–157 (2006): 3–17.

Lohrmann, Dietrich. "Idee und Wirklichkeit des Perpetuum mobile im Mittelalter." *Technikgeschiste* 73 (2006): 227–251.

Marrone, Steven. *A History of Science, Magic and Belief from Medieval to Early Modern Europe.* London and New York: Palgrave-Macmillan, 2015.

Matus, Zachary. "Reconsidering Roger Bacon's Apocalypticism in Light of His Alchemical and Scientific Thought." *The Harvard Theological Review* 105 (2012): 189–222.

Newman, William R. "An Overview of Roger Bacon's Alchemy." In *Roger Bacon and the Sciences – Commemorative Essays*, edited by Jeremiah Hackett, 317–336. Leiden: Brill, 1997.

Power, Amanda. *Roger Bacon and the Defense of Christendom.* Cambridge: Cambridge University Press, 2013.

Sannino, Antonella. *Reading William of Auvergne*. Florence: SISMEL edizioni del Galluzzo, 2022.

Sannino, Antonella. "Riforma degli studi e della società nel *Compendium studii Philosophiae* di Ruggero Bacone." *Studi filosofici* 21 (1998): 25–47.

Weill-Parot, Nicolas. "Astral Magic and Intellectual Changes (Twelfth-Fifteenth Centuries): Astrological Images and the Concept of 'Addressative' Magic." In *The Metamorphosis of Magic From Late Antiquity to Early Modern Period*, edited Jan Bremmer and Jan R. Veenstra, 167–187. Leuven: Peeters, 2002.

Weill-Parot, Nicolas. "Astrology, Astral Influences, and Occult Properties in the Thirteenth and Fourteenth Centuries." *Traditio* 65 (2010): 201–230.

Weill-Parot, Nicolas. "Causalité astrale et 'science des images' au Moyen Âge. Éléments de Réflexion." *Revue d'histoire des sciences* 52/2 (1999): 207–240.

Weill-Parot, Nicolas. "Des instruments de musique concurrents de la magie: Roger Bacon et le pouvoir de la musique." *Cahiers de Recherches Médiévales et Humanistes* 39/1 (2020): 221–235.

Weill-Parot, Nicolas. "Encadrement ou dévoilement: l'occulte et le secret dans la nature chez Albert le Grand et Roger Bacon." *Micrologus* 14 (2006): 151–170.

Weill-Parot, Nicolas. "The Magical Powers of Carved Stones from Albert the Great to Camillo Leonardi." In *Steinformen: Materialität, Qualität, Imitation*, edited by Isabella Augart, Maurice Sass, and Iris Wenderholm, 255–266. Berlin: De Gruyter, 2019.

Weill-Parot, Nicolas. *Points aveugles de la nature: la rationalité scientifique médiévale face à l'occulte, l'attraction magnétique et l'horreur du vide (xiiie-milieu du xve siècle)*. Paris: Les Belles Lettres, 2013.

Weill-Parot, Nicolas. "Roger Bacon et l'utopie scientifique au Moyen Âge." In *Les Utopies scientifiques au Moyen Âge et à la Renaissance*, edited by Roberto Poma and Nicolas Weill-Parot, 45–96. Florence: SISMEL-edizioni del Galluzzo, 2021.

Weill-Parot, Nicolas. "*Secretus* et *occultus*: lexique et modalité de l'occultation dans la science médiévale (Albert le Grand, Roger Bacon, Henri de Langenstein)." *Arcana naturae* 1 (2020): 1–20.
Weill-Parot, Nicolas. "Silence de l'auteur ou de la nature ? L'occulte et le secret dans la science médiévale (XIIIe-XVe siècle)." In *Un temps pour tout. Études sur les mutations de l'autorité de l'Antiquité au XXIe siècle*, edited by Marie Blaise and Anita Gonzalez-Raymond, 83–91. Montpellier: P.U. de la Méditérannée, 2019.
Weill-Parot, Nicolas. *Le Vol dans les airs au Moyen Âge: Essai sur une utopie scientifique*. Paris: Les Belles Lettres, 2020.
Weill-Parot, Nicolas. "Albertus Magnus, *De mineralibus, II.3.3*". In *Stein: Eine Materialgeschichte in Quellen der Vormoderne*, edited by Iris Wenderholm with collaboration of Isabella Augart. Berlin and Boston: Walter de Gruyter, 2021 pp. 191–202.
Zambelli, Paola. *The Speculum Astronomiae and Its Enigma*. Dordrecht: Kluwer, 1992.

8 Alchemy and Pharmacology in the *Opus maius*

Meagan S. Allen

The *Opus maius* is the earliest known work in which Bacon discusses the possibility of prolonging human life. It is also, in many ways, the most dissimilar to his other works on medicine. Rather than relying on alchemy, the *prolongatio vitae*, or extreme prolongation of life, is a facet of *scientia experimentalis*. In fact, the *Opus maius* provides little role for alchemy within the *prolongatio vitae*; alchemy, while also a subfield of *scientia experimentalis*, is reduced to the production of purer forms of gold.[1] This differs significantly from the *prolongatio vitae* found in the *Opus minus* and *Opus tertium*, in which alchemy teaches not only about the transmutation of metals but also the creation of a *corpus equale* – a body in which all four elements are found in equal proportion, and thus have no action upon each other.[2] The *corpus equale* is incapable of undergoing corruption, and can therefore act as a medicine for both metals and people, bringing both closer to a state of equality. In this chapter, I will characterise the nature of pharmacology in the *Opus maius*, exploring how it differs from Bacon's other medical works. Ultimately, I will argue that Bacon uses these specific pharmaceuticals because they are traditional remedies that are not alchemical in nature. He wants to give Clement IV a taste of what the *prolongatio vitae* can offer, without exposing the true alchemical nature of the *prolongatio vitae*, which he will reserve for the *Opus minus* and *Opus tertium*.

8.1 The Nature of the *Prolongatio Vitae* in the *Opus maius*

In the section dedicated to the second prerogative of *scientia experimentalis*, Bacon explains that a great tragedy has befallen the human race. At the beginning of the world, mankind lived for centuries – this was the natural lifespan of man. However, life has significantly shortened since the creation, as a result of mankind's sins. People have neglected to pay attention to the regimen of health, which weakens their bodies. These errors are compounded as the generations pass, so that fathers beget sons who are even weaker than they are. Additionally, our compounding sins have weakened our souls as well as our physical bodies, so we are no longer in full control of our bodies. As these errors have grown over the generations, we have reached our current lifespan of 70–80 years, the absolute minimum for mankind.[3] As a result, we

DOI: 10.4324/9781003405702-9

suffer from the accidents of old age: white hair, wrinkled skin, excess mucus and phlegm, injury to the sense organs, diminution of the blood, weakness of the body, forgetfulness, and failure of the natural powers of the soul. Depending on how well an individual has followed the regimen of health and the strength of their morals, the accidents of old age might set in earlier or later in life. A man who paid no attention to the regimen of health and lived a sinful life may expect to be struck down by the accidents of old age in the prime of his life, around 45–50 years.[4]

Despite our physical deterioration, Bacon says, this situation is remediable. This shortening of life was not mandated by God or nature: quite the opposite. Rather, it was accidental, a result of human shortcomings, which, therefore, meant it could be remedied using human sciences, such as *scientia experimentalis*. Nor was this just a theoretical extension of life – Bacon says that there are many authors who have been able to extend their lives using "secret experiments."[5] To be clear, Bacon is in no way advocating for the possibility of human immortality in this life. While he is certain that every human will eventually receive an immortal, and therefore incorruptible, resurrection body, this will only happen after the Last Judgement. Death is the punishment given to mankind for sin, and even though the resurrection is promised to us, everyone must experience death first. Instead, what Bacon is saying is that there are two limits to life. The first limit has been set by the conditions into which we are born – the compounding deterioration of our bodies passed down by our ancestors, and our own inclinations towards sin and the regimen of health. The second limit is the limit to human life set by God and nature. While it is possible to pass the first limit, it is impossible to pass the second.[6]

These remedies come not from the medical art – physicians are just as likely to fail in the regimen of health as anyone else – but from philosophers and experimenters.[7] Here we find a characteristic feature of Bacon's writings – the reliance upon Aristotle and natural philosophy over Galen and medical theory, even in cases directly concerning human health.[8] In his *Metaphysics*, Aristotle wrote that no corruption could occur when the active potencies of the elements are equal. All generation and corruption were caused by the action of the elements upon each other, but if the elements were to be found in equal strength – that is, their active potencies were equal – then there could be no action of one upon another, and thus no generation or corruption. In other words, an equality of potencies of the elements would lead to a state of stasis.[9] Bacon says that this mixture, in which all elements are unable to act upon each other, would form the perfect medicine. Bacon calls this a *corpus equale*, a body of equal complexion.

Because of the great difficulty involved in this process, however, few have attempted it, and even fewer have succeeded in producing a medicine capable of extending human life.[10] The *Opus maius* contains several accounts of men who were successful, but their results were proportional to the balance of the elements. Bacon recounts the story of a farmer who drank a mysterious liquor from a golden urn found in a field. This liquor had a better balance of the elements than ordinary food and drink, but did not come close to absolute perfection. For this reason, the farmer only lived for an additional 60 years. The greatest experimenter, Artephius, was able to make a medicine that had a much better balance of the elements, and so lived for 1,025 years.[11] Additional stories in the *Opus maius* include a tale from Pliny about a man who lived during the time of the Emperor Augustus, who was able to prolong his life by 100 years through applying "oil on the outside and mead on the inside"; and a story from the popular thirteenth-century medical work *De retardatione accidentium senectutis* about a man who lived for 500 years after taking a medicine meant for a king. According to Papal letters confirming this man's story, the king had requested a medicine to prolong his life. Suspicious – or perhaps doubting the medicine's efficacy – the king had it tried on a prisoner first, the celebrated individual who went on to live for another half a millennium.[12]

What is important to note about these stories is that they are coming from well-established sources of authority, and none would have been considered controversial.[13] Even though *De retardatione accidentium senectutis* was a relatively recent work, it had acquired the status of authority in medicine by the time Bacon was writing the *Opus maius*.[14] Many thirteenth-century writers, both physicians and natural philosophers, were interested in the possibility of prolonging life, and stories of people who had been able to extend their lives are not unique to Bacon.[15] Likewise, the majority of the medicines Bacon writes about in the *Opus maius* can best be described as "conventional."

Before exploring the types of medicines described in the *Opus maius*, it is important to recognise that Bacon is writing for a specific purpose – to show that the *prolongatio vitae* is possible. He wants Clement IV to understand that the shortening of human life is accidental and can be remedied. The majority of the section of the *Opus maius* on the *prolongatio vitae* is devoted to the causes of human deterioration and their symptoms, and stories of men who have successfully prolonged their lives – only a few paragraphs are devoted to specific medicines. What Bacon is not doing is writing a manual on how to prolong life. The *Opus maius* will not teach anyone how to make the medicines that will extend their life; that is not Bacon's goal. He is not a physician or apothecary; he will not teach how to properly compound the medicines that prolong life. Rather, he is simply showing us that such prolongation is a possible and noble goal.[16]

The first "medicine" Bacon describes are those things used by animals to prolong their own lives, which humans should take as examples. Certain animals, such as snakes, deer, and eagles, were considered to have extremely long lifespans. Despite not having the capacity for reasoning, these animals were given knowledge by God as to what stones and herbs to ingest so that they would have long lives. The wisest men know that we should use these animals as examples, eating the stones and herbs that they do to prolong our own lives. We know this not just from our own experiments, Bacon says, but from the works of the ancients, including Pliny, Solinus, and Tully.[17] The second medicine, mentioned only in passing, is called the *gloria inestimabilis*, and tempers the humors to preserve health and delay ageing.[18]

The third "medicine," and the one to which Bacon devotes the most time, is a list of seven occult medicines, which I will refer to as the "secret seven." These are seven medicines found in *De retardatione accidentium senectutis* that can prolong life.[19] Though described in vague terms in Bacon's source material, the *Opus maius* explicitly states the identity of the seven: gold, pearls, rosemary, ambergris, lign-aloes, the bone from the heart of a stag, and the flesh of the Tyrian viper, from which theriac is made.[20]

Like the stories of men who prolonged their lives by centuries, there is nothing new about the secret seven. Despite their rarity or cost to the average individual, all seven are standard medicines found in many ancient and medieval medical compendia. The idea that there are health properties associated with a bone found in the heart of a stag, for instance, can be seen as early as Hesiod.[21] Importantly, there is nothing objectionable to any of the medicines Bacon describes in the *Opus maius*. Even if they act by means of occult properties that cannot be fully understood, almost all authorities are in agreement that they do have some beneficial effect on the human body. That a medicine worked by an occult property was not a reason to discard it; it merely meant that its effects had to be determined experimentally, rather than be reasoned out.[22] Even the most conservative physicians could have found nothing to fault in Bacon's description of the secret seven.

8.2 The *Opus maius* in Bacon's Oeuvre

Bacon's programme of the *prolongatio vitae* was not solely contained in the *Opus maius*. Rather, it was something that continued to fascinate him from the time he wrote the *Opus maius* until his death in 1292 – a period of almost 30 years. However, there are two important characteristics that set the *Opus maius* apart from almost every other genuine work in Bacon's oeuvre: the denial of the role of alchemy in medicine and the reliance on the secret seven as a body of medicines.

In the centuries after his death, many works on alchemy and medicine were attributed to Bacon, including the aforementioned *De retardatione accidentium senectutis*. However, recent scholarship has shown that many of these works are spurious.[23] Currently, there are eight medical/alchemical texts that can be attributed to Bacon with a high degree of certainty: *Antidotarius*, *De erroribus medicorum*, *Epistola de secretis operibus artis et naturae et de nullitate magiae*, *Liber sex scientiarum*, the three *Opera*, and his introduction and glosses on the Pseudo-Aristotelian *Secretum secretorum*.[24] Any description of Bacon's programme of the *prolongatio vitae* must rely on these works.

With the obvious exception of the *Opus maius*, Bacon's alchemical theories remain largely consistent throughout his writings. Alchemy is the key to medicine because alchemy teaches about the generation and corruption of all materials in the sublunary world.[25] According to Bacon, there are three "layers" to the material world. The first layer consists of the four elements: fire, air, water, and earth. Borrowing from his classical sources, Bacon says that these four elements combine in different proportions to form different substances. He calls these substances the simple humors: blood, phlegm, bile, and melancholy. Each of the four simple humors is dominated by one element, which gives it its characteristics and qualities (in the Aristotelian sense). Blood, for example, is predominantly air, a hot and moist element. Bile, on the other hand, is predominantly fire, which is a hot and dry element. In the following manner, each of the simple humors is formed:

FIRE (Hot and Dry) + some air + some water + some earth = Bile
AIR (Hot and Wet) + some fire + some water + some earth = Blood
WATER (Cold and Wet) + some fire + some air + some earth = Phlegm
EARTH (Cold and Dry) + some fire + some air + some water = Melancholy

The simple humors thus constitute the second "layer" of the material world. Much like the four elements combine in different proportions to form the four simple humors, the four simple humors themselves form mixtures of different proportions to create the four compound humors. Bacon refers to the compound humors by the same names: blood, bile, phlegm, and melancholy. Once again:

SIMPLE BILE (Fire) + simple blood + simple phlegm + simple melancholy = Compound Bile
SIMPLE BLOOD (Air) + simple bile + simple phlegm + simple melancholy = Compound Blood
SIMPLE PHLEGM (Water) + simple bile + simple blood + simple melancholy = Compound Phlegm
SIMPLE MELANCHOLY (Earth) + simple bile + simple blood + simple phlegm = Compound Melancholy

When the physicians speak of the four humors in a person's body, Bacon says, they are referring to the compound humors. However, these materials are found in all things, not just animate ones. They are only properly called "humors" when found in humans or animals, but they are the same material that all things in the sublunary world are generated from – plants, animals, or minerals. Alchemy is the science that teaches about how these layers are created and destroyed, and thus about the generation and corruption of all sublunary materials, as well as their interactions with each other.[26] Because the art of medicine is based on the principle of balancing the humors, physicians ought to know how the humors are formed, but also how they will respond to the various medicines that are being used to bring the humors back into balance.[27]

Unfortunately, Bacon says, there are few physicians who learn the principles of alchemy, for it is not something that is taught in the medical curriculum.[28]

Alchemy's utility to the physician cannot be overstated. Perhaps the most obvious way is by taking common medications and making them easier and safer to use. *De erroribus medicorum* is filled with examples of popular medicines that have potentially dangerous side effects, including opium, snake flesh, and rhubarb. An incorrect dose of one of these might have disastrous consequences for the patient. Instead, Bacon cautions that the physician must prepare the medicines alchemically to remove their potentially dangerous aspects. Snake flesh, for example, was thought to contain poisonous material, which might injure the patient if administered incorrectly. By preparing the medicines alchemically, one could remove the poisonous material, leaving only the beneficial substance behind. This could then be given to the patient without concern as to any side effects.[29]

Even medicines that did not have any poisonous qualities could benefit from alchemical preparation. Bacon is particularly concerned about the ingestion of stones and metals.[30] He warns that there is little therapeutic value to swallowing such stones whole, as they are often done, because they are passed out of the body without imparting their desired effect. Essentially, the body is unable to digest the stone in its whole form, and the patient receives no benefit. Instead, Bacon claims, these stones ought to be calcined and ground into a fine powder. This would save essential time and energy during the digestion process, enabling the body to make use of the therapeutic aspect of the stones.[31]

Even more importantly, alchemy taught about the creation of the *corpus equale* – that body of equal complexion in which there was perfect stasis from an inability of the elements to act upon each other. Because alchemy alone taught about the generation of all things from the four elements, it likewise taught how to properly balance the elements in any given material to bring them closer to this state of perfection. The skilled alchemist could separate and purify the elements in any material, then recombine them in the proper proportion to make a *corpus equale*.

In fact, Bacon claims, there are three things that have this *corpus equale* in the sublunary world: gold, the eventual resurrection body, and materials that have had their elements properly balanced by the alchemist. This *corpus equale* would then serve as a medicine, passing its equality on to the recipient body through the multiplication of its species of equality. It could cure any disease, prolong life, and even transmute base metals into noble ones.[32] Bacon is very committed to this idea – a form of this statement highlighting the importance of alchemy can be found in all of the aforementioned works with the exception of two – the *Opus maius* and the *Antidotarius*.[33]

While the idea of the *corpus equale* and the *prolongatio vitae* certainly does appear in the *Opus maius*, it is connected to *scientia experimentalis*, not alchemy. Alchemy is confined only to the creation of metals. The purest gold that nature can produce has 24 "grades" or "degrees." Like nature, alchemy can also produce gold of 24 grades, but cannot produce anything purer. *Scientia experimentalis*, however, can produce an even purer gold than nature can – up to 30 or even 40 degrees. The explicit connection between the medicine for people and the medicine for metals already exists in Bacon's mind, but again, falls under the domain of *scientia experimentalis*, not alchemy.[34] Alchemy is the foil to *scientia experimentalis* – where alchemy fails, *scientia experimentalis* succeeds. However, by the time he writes the *Opus minus*, alchemy is a necessary part of natural philosophy, alongside mathematics and *perspectiva*. Bacon explains that:

> Alchemy is experimental, and with mathematics and perspective profit of the body and of [its] condition. For towards the whole of health and the prolongation of life by the secrets of nature will these sciences be sufficient, and it completes the science of ordinary medicine, and it adds utmost advantages of bodies.[35]

This sentiment is repeated, even more strongly, in the *Opus tertium*, where Bacon writes, "Therefore, because of ignorance of this science [alchemy], it is not possible to know of conventional natural philosophy, nor speculative medicine, nor by consequence practical medicine."[36]

What, then, are we to make of this? Has Bacon changed his mind on the role of alchemy? Has he had some personal experience that suggests that alchemy is far more powerful than he previously thought? One section on alchemy and the *prolongatio vitae* in the *Opus minus* contains a bizarre, but potentially illuminating, reference to a female mystic living in the diocese of Norwich who went 20 years without eating. While abstaining from eating, she did not pass any excretions from her body, and remained in good health, as confirmed by the bishop. This was not a miracle, but rather a work of nature – the woman's body had been reduced to a state closer to equality by a beneficent constellation. Surely such a case would provide Bacon with much-needed proof for his theory of the *prolongatio vitae* based on a *corpus equale*.[37]

However, Bacon does not use this story for anything else in the *Opera*, not even as part of a long discussion on the process of digestion in the *Opus tertium*.[38] This should not be surprising – not only is Bacon writing in a system in which case histories have yet to become the norm in medical literature, he is also part of a system of natural philosophy in which nature behaves according to strict rules. The elements can only act in certain ways upon each other, and if they are balanced in potency, then there can be no action. This is according to the laws of God and nature, and Bacon, nor any other medieval scholastic, would need to see an individual case for proof of this rule.

Bacon himself acknowledges this discrepancy in both the *Opus minus* and *Opus tertium*, and explains his apparently negative attitude towards alchemy as follows:

> And it pleases me to discuss these things in this place in order that the secrets of alchemy may be concealed in these things which I write, but that you may observe them with discretion. Unless it be known what I wrote earlier, and that which I will write, and thirdly that which I will insert at the end of the work, it is impossible to know the truth of these secret writings.[39]

A similar passage specifically touching upon alchemy and medicine appears in the *Opus tertium*:

> Therefore I wrote of these secrets in three places to Your Grace. For in the second work, I wrote first of practical alchemy under enigmas, in the manner of philosophers, acknowledging their enigmas, and inserting some others following the fact that the latter use the same liberties as the former. Then, after much writing, I wove together all the roots of speculative alchemy into the sixth sin of the study of theology; and there is contained much which is influential to the understanding of the former, but is not sufficient. I sent out the writing from my own hand through John, in order that it be copied for Your Glory; and there, although there is great concealment, still it is not through enigmatic words, for I explain many of them, but it is through a wise and philosophical method. For since natural philosophy, medicine, and alchemy share common roots, I therefore pretended to give these roots as if they were only natural and medicinal, when they are really alchemical, and I introduced them as such.[40]

Bacon has two reasons for splitting the information on alchemy throughout three works: first, he wants to protect the information that he is sending to Clement IV – fearful that it may fall into the wrong hands, he splits it up. As a particularly potent science, it was especially important that alchemy's secrets be protected from those who would use it to do evil. If one work should fall into the wrong hands, Bacon can rest easy that they will not be able to truly understand the secrets of alchemy without the other two. Second, Bacon also sees both the *Opus minus* and *Opus tertium* as opportunities to expand upon the theories proposed in the *Opus maius*. Although he did not have the time to sufficiently deal with alchemy and the *prolongatio vitae* in the *Opus maius*, he will do so in the *Opus minus* and *Opus tertium*.[41]

8.3 The Secret Seven and the *Opera*

Pharmacology in the *Opus maius* can essentially be reduced to the secret seven. Bacon describes them as follows:

> Therefore the excellent experimenter in the book on the Regimen of the Aged says that if what is tempered in the fourth degree, and what swims

> in the sea, and what grows in the air, and what is cast up by the sea, and a plant of India, and what is found in the vitals of a long-lived animal, and the two snakes which are the food of the Tyrians and Aethiopians, be prepared and used in the proper way, and the minera of the noble animal be present, the life of man could be greatly prolonged and the conditions of old age and senility could be retarded and mitigated.[42]

But Bacon is very candid about their true natures. In the next sentence, he explains that:

> But what is tempered in the fourth degree is gold [....] And if by a certain experiment gold should be made the best possible, or at any rate far better than nature and the art of alchemy[43] can make it ... it would then produce a wonderful action on the body of man. And if there is added that which swims in the sea, namely, the pearl, which is a thing most efficacious for preserving life, and there is added also the thing that grows in the air. This last is an Anthos and is the flower of seadew, which possesses an ineffable virtue against the conditions of old age [...] to these must be added what is cast up by the sea. This last is ambergris, which is spermaceti, a thing of wondrous virtue in this matter. The plant of India is similar to these, and is the excellent wood of the aloe [....] To these ingredients there is added that which is in the heart of a long-lived animal, namely, the stag. This is a bone growing in the stag's heart, which possesses great power against premature age. The snake which is the food of the Tyrians is the Tyrian snake from which Tyriac is made [....] The Snake that is the food of the Aethiopians is the dragon [...] and they use the flesh against the accidents of old age, and they prolong life and sharpen their intellect beyond all conception.[44]

Bacon is quite clear that these remedies do not originate with him, but come from *De regimine senum* and *De retardatione accidentium senectutis*, which were themselves borrowing from a long tradition of medical literature.[45] We should not consider Bacon to be innovative in his inclusion of any of these. Rather, he is engaging with a much larger medical tradition that contained certain "stock" pharmaceuticals.

Nor are their actions unusual for thirteenth-century medical theory: gold, pearls, and the bone from the heart of a stag all act by preserving both the natural heat and the radical humidity of the body. With these two things in balance, the author of *De retardatione accidentium senectutis* claims that no one needs to fear sickness. Interestingly, the author also considers the idea that the medicines can act as a *corpus equale* – that is, a substance that is itself prone to preservation will preserve another thing, just as something that is prone to corruption will corrupt another thing. However, this should not necessarily be read alchemically – the author of *De retardatione accidentium senectutis* gives no indication that he is connecting this to any alchemical theory. The preservative properties of certain substances, such as alcohol, were a popular topic of discussion in thirteenth-century medical literature.[46]

Given their importance in the *Opus maius*, we might expect Bacon to include the secret seven in the *Opus minus* and *Opus tertium*. However, they do not appear. In fact, the only time that they appear as a whole again in Bacon's entire oeuvre is in *Corpora adae et evae*, which is itself almost a word-for-word copy of the section on experimental science in the *Opus maius*. There are scattered references to individual members of the secret seven, and while one or two may be grouped together, all seven never appear in the same place in any other novel work by Bacon. This has led some scholars to suggest that Bacon may have rejected *De retardatione accidentium senectutis* and *De regimine senum* as medical authorities after completing the *Opus maius*.[47]

In this scenario, Bacon would reject a previously authoritative medical work, which relies on humoral theory and the actions of innate heat and radical humidity to explain ageing, and replace it with a more "alchemical" work, which taught of a *corpus equale*, such as Pseudo-Avicenna's *De anima*.[48] Again, while it is possible that Bacon began to place more importance on alchemical works after completing the *Opus maius*, I do not think this is a wholesale rejection of one form of literature in favour of another. Rather, it seems evident that Bacon has always intended to have one idea spread across three works. At the very least, this sense of unity is what he wishes to portray in the *Opus tertium* and in his letter to Clement IV.

Is there anything else that the appearance – or lack thereof – of the secret seven in Bacon's works tells us about his theories of the *prolongatio vitae*? The remaining two works in the *Opera* do not mention any of the secret seven. The *Opus minus* focuses almost entirely on the *corpus equale* and the state of the human body in Paradise and after the resurrection.[49] The section on the *prolongatio vitae* in the *Opus tertium* begins in much the same way as in the *Opus maius* – Bacon begins by stating that while it is possible to prolong life, this is not something that is taught in the books of medicine. Experimenters have been able to figure out this science by following the examples of brute animals, such as eagles, deer, and serpents, who have been given knowledge of how to prolong their lives. But in addition to the knowledge gained by copying the animals, these men have figured out how to use practical alchemy to prepare a body of equal complexion to prolong their lives.[50] Has Bacon thus rejected the secret seven because they do not operate through a *corpus equale*? He is now certain that this is how a medicine that will prolong life must act, for this was the state of man's body in innocence and will be the state of the body after the resurrection.

The *Opus maius*, *Opus tertium*, *Secretum secretorum*, Bacon's other novel medical works: the *Antidotarius*, *De erroribus medicorum*, and his glosses on the *Secretum secretorum* contain various references to some of the secret seven.[51] The *Epistola de secretis operibus artis et naturae et de nullitate magiae*, which, for the purposes of this chapter, I have chosen to regard as authentic, contains no mention of any of the secret seven, despite having an entire chapter devoted to the *prolongatio vitae*. This chapter contains many of the same stories as the *Opus maius*, and some sections are repeated almost verbatim.[52] Likewise, the *Liber sex scientiarum*, which extolls the power of alchemy, astronomy, and perspective in prolonging life, makes only a passing mention of gold. Following in the same vein as the *Epistola*, the *Liber sex scientiarum* repeats many of Bacon's favourite stories of men who have prolonged their lives by following the examples of brute animals, including the legendary Artephius. Bacon says that wise men like Artephius can use experiments to concentrate stellar rays to make medicines more efficacious, but the masses who use gold for diseases like leprosy don't bother to do this, and so the gold cannot assimilate to human nature.[53] It appears that there is roughly a reverse correlation between the use of alchemy and the secret seven in Bacon's works – those works that rely heavily on alchemical principles or mention alchemy by name do not incorporate the secret seven to the same extent as the works which do not prioritise alchemy. The only outlier is *De erroribus medicorum*, which includes both six of the seven occult remedies and alchemical theory.

8.4 Conclusion: Why Do the Secret Seven Disappear?

In the late thirteenth century, the General Chapter of the Franciscan Order banned the study, practice, and even possession of alchemical works among its members. The punishments for transgressors could be quite severe, including imprisonment and exhumation (if the member was deceased).[54] While there are fourteenth-century records suggesting Bacon was imprisoned for some form of "suspected novelty," whether or not he was actually imprisoned by the order has never been confirmed by contemporary documentation.[55] It seems unlikely that he would actually have been imprisoned, even given his firm belief in alchemy. Nor were any of his other sciences: *perspectiva*, astronomy, etc., particularly "suspicious."[56]

In the *Opus maius*, pharmacology and the *prolongatio vitae* are essentially reduced to the secret seven – seven medicines, typical of thirteenth-century pharmacopoeias, that were found grouped together in the popular *De retardatione accidentium senectutis*. There is no alchemical basis behind their effectiveness, and alchemy itself is a negative foil to the much better *scientia experimentalis*. Even the *corpus equale*, while also important for the *prolongatio vitae*, is treated separately from the secret seven and is not given as the rationale for their efficacy. Rather, they seem to work via some occult property that Bacon takes for granted because it has already been determined experimentally. Likewise, the stories of men who prolong their lives to extreme lengths do not rely on the secret seven but on bodies of equal complexion.

As early as the *Opus minus* and *Opus tertium*, Bacon has rejected this scheme. Alchemy is fundamental to learning about – and making – the *corpus equale*, which is the best medicine. We can conclude from the rest of

Bacon's oeuvre that this is his true belief about alchemy – it is not a failed version of *scientia experimentalis*, but rather, the basis of natural philosophy. Why then the subterfuge? Has Bacon drastically changed his mind in the months between completing the *Opus maius* and the *Opus minus*? While it cannot be ruled out that Bacon experienced some sort of confirmation of the powers of alchemy, such as the female mystic suffering from anorexia mirabilis, or read a new translation of an Arabic alchemical work like Pseudo-Avicenna's *De anima*, it seems more likely that this is a case of him deliberately splitting up his ideas on alchemical pharmacology over the *Opera*. As William R. Newman has shown in this volume, Bacon was a master of alchemical riddles, incorporating different puzzles and enigmas into his writing to disguise his true intentions from those whom he deemed unworthy of the knowledge.[57]

I propose that Bacon uses the secret seven in the *Opus maius* because they are a convenient placeholder. This is the work in which he says the least about the powers of alchemy, and so it makes sense that it would have no "alchemical" medicines. The secret seven are not alchemical – they are traditional remedies that have been used for centuries in European medicine and contain nothing objectionable. Once the *Opus maius* is completed, however, Bacon no longer needs to have a "placeholder" – he can instead openly discuss the medicines that he believes will really prolong life – those made of an equal complexion. The *Opus maius* is the only work in which Bacon explicitly denies the powers of alchemy, and it is the only work in which the secret seven appear as a group.

The question then becomes – to what extent was Bacon's compulsion to divide his alchemical remedies over three works driven by fear of censorship versus fear of this precious information falling into the wrong hands? It is certainly true that Bacon feels that he has to complete the *Opera* in secret, without the knowledge of his superiors. But if this is his reasoning, then why risk including alchemy at all in either the *Opus minus* or *Opus tertium*? Perhaps it is the case that despite the multiple acts banning alchemy in the Franciscan Order during the thirteenth and fourteenth centuries, Franciscan authorities were more forgiving of the alchemical art than their laws would let on. We know that there were a number of alchemists with ties to the Franciscan Order in this period.[58] It seems more likely that while Bacon may have been afraid of censorship from his direct supervisors at Oxford or Paris, he was probably more worried about his knowledge being shared with those whom he deemed unworthy.

Notes

1 Bacon, *Opus maius*, vol. II, Part 6. English translation in *The Opus Majus of Roger Bacon*, 2 vols, translated by Robert Belle Burke. References to the *Opus maius* will be to Bridges' edition with Burke's translation in parentheses. For more on *Scientia Experimentalis*, see the chapter by Jeremiah Hackett in this volume.

2 Bacon, *Opus tertium,* 878.
3 Bacon, *Opus maius*, vol. II, 205 (618). Bacon takes the limit of 70–80 years from Psalm 90. As for the regimen of health, Bacon equates that with following the six non-naturals, generally defined as food and drink, motion and rest, sleep and wakefulness, elimination and retention, air, and passions and emotions. See Bacon, *Opus maius*, vol. II, 205 (617).
4 Bacon, *Opus maius*, vol. II, 206 (619).
5 Bacon, *Opus maius*, vol. II, 208 (620).
6 Bacon, *Opus maius*, vol. II, 207 (620). Bacon claims this idea comes from Scripture, which states "Though hast set its bounds which cannot be passed."
7 Bacon, *Opus maius*, vol. II, 207 (620). Bacon says that many men have written about such prolongation of life, including Dioscorides, Hali, and Avicenna.
8 Allen, *Roger Bacon and the Incorruptible Human,* Chapter 6.
9 This is also the state in which our bodies will exist after the resurrection, and the reason why we will be immortal. As mixed bodies, our bodies desire this equality and the subsequent fulfilment of the appetite of matter. Adam had a near-perfect equality of elements, and he would have lived forever in Eden. The fruit of the tree of life was a *corpus equale* and would have sustained Adam's body forever. Bacon, *Opus maius*, vol. II, 212 (625).
10 Bacon, *Opus maius*, vol. II, 213 (625). Bacon takes this from Aristotle's *Metaphysics*, but also says that a similar position is held by Averroes, in opposition to Galen.
11 Bacon, *Opus maius*, vol. II, 209–210 (625). According to *De retardatione accidentium senectutis*, this man lived during the time of King William of Sicily. One day while ploughing the fields, he found a golden vessel containing a mysterious liquor. Thinking it was dew, he drank the liquor and washed his face, and his youth was renewed. He lived for another 60 years and served as William's messenger. Bacon also includes other stories from *De retardatione accidentium senectutis*, including that of a man who used an unguent to anoint his whole body and lived for several centuries without any pains except on the soles of his feet, where he had neglected to anoint himself.
12 Bacon, *Opus maius*, vol. II, 210 (622).
13 Allen, *Roger Bacon and the Incorruptible Human*, 64–70.
14 For many years, it was incorrectly believed that Bacon was the author of *De retardatione accidentium senectutis*. However, this is highly unlikely. See Bagliani, "Ruggero Bacone autore del *De retardatione accidentium senectutis,*" 707–728. *De retardatione accidentium senectutis* heavily borrows from Haly Abbas' *Liber totius medicine necessaria*, which had been translated into Latin in 1127. See Bacon, *Opus maius*, vol. II, 206, footnote 1.
15 Gruman, "A History of Ideas about the Prolongation of Life," 1–102.
16 Although he does not say it explicitly in the *Opus maius*, Bacon does state in *De erroribus medicorum* that his purpose is not to instruct but to arouse curiosity. The same is almost certainly true of the *Opus maius*, given its purpose as a plea for reform to Clement IV.
17 Bacon, *Opus maius*, vol. II, 208 (620–621).
18 Bacon, *Opus maius*, vol. II, 208–209 (621). Bacon takes the *gloria inestimabilis* from the Pseudo-Aristotelian *Secretum secretorum*, which contains a recipe and instructions for its preparation. See Bacon, *Secretum Secretorum*, 101–104.
19 In the *Opus maius*, Bacon references *De regimine senum*, not *De retardatione accidentium senectutis* (*Libro de accidentibus senectutis*), as the source of the secret seven. Bridges identified *De regimine senum* as Part 2, Book 1, Chapter 24 of Haly Abbas' *Liber totius medicine necessaria*. However, references to these occult remedies do not appear in this section of Haly Abbas' work. They do, however,

appear in great detail – and in strikingly similar terms to the ones Bacon uses – in *De retardatione accidentium senectutis*. As I have already demonstrated, *De retardatione accidentium senectutis* was heavily reliant upon the *Liber totius medicine necessaria*, and the fact that the occult remedies do not appear in this chapter does not mean that they are not present at all in Haly Abbas' work. Bacon certainly read both *De retardatione accidentium senectutis* and the *Liber totius medicine necessaria*, for he makes direct references to their contents in other places. Ultimately, for the sake of this argument, it does not matter whether Bacon found the secret seven in the *Liber totius medicine necessaria* or *De retardatione accidentium senectutis*. What matters is that he was borrowing them from an established authority, which both sources were.

20 Bacon, *Opus maius*, vol. II, 210 (622–623).

21 Little, "Introduction to *De retardatione accidentium senectutis*," xli.

22 McVaugh, "Theriac at Montpellier 1285–1325," 113–144.

23 Paravicini Bagliani, "Ruggero Bacone autore del *De retardatione accidentium senectutis*," 707–728; Newman, "The Alchemy of Roger Bacon," 461–479; and Allen, *Roger Bacon and the Incorruptible Human*, Appendix B. Following Agostino Paravicini Bagliani's lead, I have omitted the three short medical works attached to *De retardatione accidentium senectutis: De balneis senum et seniorum, De universali regimine senum et seniorum*, and *De compositione quarundam medicinarum*. Though these works are usually attributed to Bacon, they are found appended to *De retardatione accidentium senectutis*, shedding doubt on their authenticity. There are two further works that deserve closer inspection to determine authenticity: *Corpora adae et evae* and *In debito regimine*. For more on these works, Tavormina, "Roger Bacon." Tavormina has rightly pointed out that *Corpora adae et evae* is almost identical to the *Opus maius*, and so for the purposes of this chapter, I do not consider it to be a novel work – that is, a work in which Bacon expresses any original ideas not found in any of his other writings. It also contains sections that are copied almost verbatim from the *Opus minus*. Likewise, *In debito regimine* is almost identical to the *Liber sex scientiarum* in many places.

24 While it cannot be determined conclusively that the *Epistola de secretis operibus artis et naturae et de nullitate magiae* is a work by Bacon, I have argued elsewhere that the evidence currently supports the theory that it is, and that it should therefore remain within his oeuvre. See Allen, *Roger Bacon and the Incorruptible Human*, Appendix C.

25 Like many of his sciences, Bacon divides alchemy into practical and speculative parts. Speculative alchemy teaches about the generation and corruption of materials from the elements, while practical alchemy teaches how to make precious metals from base ones and medicines that can extend human life. Of the two, medicines that can extend human life are far more important to Bacon. See Bacon, *De erroribus medicorum*.

26 Bacon, *De erroribus medicorum*, 159–160.

27 Medieval pharmacology was largely based on a system of antipathy, in which a medicine of one quality acted on a disease of the opposite quality to bring about the desired effect. For example, a "hot" disease, such as a fever, would be treated by taking a "cold" medicine.

28 Bacon, *De erroribus medicorum*, 155.

29 Bacon, *De erroribus medicorum*, 155–156. A similar principle applied to rhubarb, which was thought to be very hot, even dangerously so. It thus had to be used with caution. Bacon is not the first to argue that the "poisonous" part of *materia medica* must be removed before it can be used – this idea is also found in *De retardatione accidentium senectutis*. However, this is not explicitly connected to alchemy. See Little, *De retardatione accidentium senectutis*, 17–20.

30 The use of stones and metals for certain diseases, especially those relating to the principal seats of the body, such as the heart, was a standard aspect of ancient and medieval medicine. Stones could be ingested, or they could also be worn on the body as talismans. For more on the use of stones in medicine, see Ziegler, *Medicine and Religion c. 1300*.

31 Bacon, *De erroribus medicorum*, 165. Essentially, Bacon is saying that by calcining and grinding the stones or metals, the alchemist has mimicked the first stage of digestion. As with consuming any predigested material, the body no longer has to expend as much energy in processing the nutritious matter from the waste.

32 Allen, *Roger Bacon and the Incorruptible Human*, Chapter 4.

33 The *Antidotarius* also contains no alchemical theory but is very closely related to other thirteenth-century pharmacopoeia. It contains descriptions of the different pharmaceuticals available to physicians and how they should be prepared. Bacon's main concern in this work is that the physician understands that certain compound medicines, especially those that contain opium, work by a power greater than the sum of their parts. This means that they must have their effects determined experimentally and must be prepared correctly in order to have these desired effects – usually through the process of fermentation. In fact, the *Antidotarius* may itself be regarded as unusual in its high degree of practicality. Though two of the secret seven do appear in the *Antidotarius*, they do so in the context of the importance of understanding the roots of medicines. Bacon says that it is crucial to understand the roots of medicines – viper flesh for theriac and aloes in ierapigra – in order to know how a compound medicine will work. One must know how to remove the poisonous parts of the root and compound it in the proper proportion with other ingredients in order to make a beneficial medicine. What he does not do is discuss how these two ingredients prolong life, other than saying that they purge all poisons from the body. See Bacon, *Antidotarius*.

34 Bacon, *Opus maius*, vol. II, 214–215 (626–627).

35 Bacon, *Opus minus*, 324. Alkymia est experimentalis, et mathematica et perspectiva bonum corporis et fortunae. Nam ad omnem sanitatem et propagationem vitae per secreta naturae sufficiunt hae scientiae, et perficiunt scientiam medicinae vulgatae, et ad eam addunt utilitates corporum maximas.

36 Bacon, *Opus tertium*, 80. Unde, propter ignorantiam istius scientiae, non potest sciri naturalis philosophiae vulgata, nec speculativa medicina, nec per consequens practica.

37 Bacon, *Opus minus*, 373–374. Et etiam aliqui diu vixerunt sine nutrimento, ut nostris temporibus fuit una mulier in Anglia in diocesi Norwicensi, quae non commedit per xx annos, et fuit pinguis et in bono statu, nullam superfluitatem emittens de corpore, sicut probavit episcopus per fidelem examinationem. Nec fuit miraculum, sed opus naturae, nam aliqua constellatio fuit illo tempore potens elementa reducere ad gradum aequalitatis propinquiorem, quam ante fuerunt in corpore suo; et quia eius commixtio fuit ex natura propria conveniens constellationi non sic aliorum complexiones, ideo accidit alteratio eius in corpore quod non in aliis. I have not been able to identify this woman. There are numerous female mystics in and around medieval Norwich, but none who were famous for anorexia mirabilis were active during or immediately prior to the time Bacon was writing the *Opus minus*. See Bynum, *Holy Feast and Holy Fast*; and Hill, *Women and Religion in Late Medieval Norwich*. The only woman I have found to fit this description is Marie d'Oignies (d. 1213), a Belgian mystic from Nivelles, who underwent prolonged periods of fasting during which she would only eat the Host. Her life was chronicled first by her friend Jacques de Vitry, Bishop of Acre, and then by Thomas of Cantimpre. However, there is no record that Marie ever travelled to Norwich, nor did she go 20 years without eating.

38 Bacon, *Opus tertium*, 320–340. This story is repeated in *Corpora adae et evae*. See Tavormina, "Two Extracts on Prolonging Life," 353.
39 Bacon, *Opus minus*, 360. Et placet mihi in hoc loco de his tractare uti secreta alkymiae occultentur in his quae scribo, quod tamen vestra discretio valeat contemplari; nisi enim sciantur quae prius scripsi, et ea quae hic scribam, et tertio ea quae in fine operis conscribo, non potest horum sciri veritas scriptorum secretorum.
40 Bacon, *Opus tertium*, 954. Scripsi enim in tribus locis Vestre Glorie de huiusmodi secretis. Nam in Secundo Opere, scripsi primo de alkimia practica sub enigmatibus, more philosophorum, innuens enigmata eorum, et inserens aliqua alia secundum quod posteriores ea libertate utuntur qua priores. Deinde, multa scriptura revoluta, texui omnes radices alkimie speculative, in 6 peccato studii theologie; et ibi continentur multa que valent ad intellectum precedentium, sed non sufficient. 3 autum scriptum misi de manu mea per Joannem, ut Vestre Glorie transcriberetur; et ibi, licet sit occultatio multiplex, tamen non est per verba enigmatica, nam illa expono multum, sed est per modum magis philosophicum et sapientialem. Nam quia communicant in radicibus naturalis philosophia, medicina et alkimia, ideo simulavi me tradere radices has tanquam essent solum naturalia et medicinalia, cum tamen sunt alkimistica, et pro talibus ea introduxi.
41 Bacon, *Opus tertium*, 136–138. This, of course, does not rule out the possibility that Bacon read or saw something else that strengthened his views on the ability of alchemy to prolong human life. Merely that this idea was already in his mind when he was writing the *Opus maius*. For more on Bacon and the practice of *dispersa intentio*, see the chapter by William Newman in this volume.
42 Bacon, *Opus maius*, vol. II, 210 (622). Et ideo dicit experimentator bonus in libro de Regimine Senum, quod si illud quod est in quarto gradu temperatum, et quod natat in mari, et quod vegetatur in aere, et quod a mari projicitur, et planta Indiae, et quod est in visceribus animalis longae vitae, et duo repentia quae sunt esca Tyrorum et Aethiopum, praepararentur et adhiberentur, ut oportet, et minera nobilis animalis adesset, multum posset vita hominis prolongari, et passiones senectutis et senii retardari et mitigari.
43 Once again, Bacon uses alchemy as an example of an inferior science, unable to match the powers of *Scientia Experimentalis*.
44 Bacon, *Opus maius*, vol. II, 210–211 (622–623). Quod vero est in quarto gradu temperatum est aurum [....] Et si per experientiam certam fieret optimum quod potest esse, vel saltem longe melius quam natura et ars alkimiae possunt facere [...] tunc miram operationem faceret in corpus hominis. Et si jungitur illud quod natat in mari, videlicet, margarita, quae est res multum efficax ad vitae conservationem, ac etiam addatur res quae in aere vegetatur, et est Anthos, et est flos roris marini, qui ineffabilem habet virtutem contra passionem senectutis [....] Quibus addendum est quod a mari projicitur et est ambra, quae est sperma cetae, res mirae virtutis in hac parte. Planta Indiae his est conformis, et est bonum lignum aloes, novum, non sophisticatum. Quibus annecitur quod est in corde animalis longae vitae, scilicet cervi, et est os quod generatur in corde cervi, habens magnam proprietatem contra festinationem senectutis. Repens autem quod est esca Tyrorum, est tyrus serpens de quo fit Tyriaca [....] Repens quod est esca Aethiopum est draco [...] sicut est ars praeparandi carnes tyri, et utuntur eis contra accidentia senectutis, et vitam prolongant et intellectum subtiliant ultra omnem aestimationem.
45 Allen, *Roger Bacon and the Incorruptible Human*, 130–132.
46 Little, *De retardatione accidentium senectutis*, 16, 42–46. For more on the use of alcohol in thirteenth-century medical literature, see Rasmussen, *The Quest for Aqua Vitae*.
47 Paravicini Bagliani, "Ruggero Bacone autore del *De retardatione accidentium senectutis*," 707–728; and Paravicini Bagliani, "Ruggero Bacone e l'alchimia di lunga vita," 51–54.

48 For more on the innate heat and radical humidity, see McVaugh, "The 'Humidum Radicale' in 13th Century Medicine," 259–283.

49 Bacon, *Opus minus*, 367–369, and Roger Bacon, *Opus minus*, in Rome, Biblioteca Apostolica Vaticana MS Reginensis Latini 1317, folios 108v–132r.

50 Bacon, *Opus tertium*, 43–44, 872–874.

51 For an explanation as to why the *Antidotarius* has been omitted from this comparison, see footnote 32.

52 Bacon, *Epistola de secretis operibus artis et naturæ.*

53 Bacon, *Liber sex scientiarum*, 183. Divites homines multotiens utuntur auro contra lepram et multas graves infirmitates, sed quia non preparatur modo dicto, ideo non incorporatur humane nature.

54 Theisen, OSB. "The Attraction of Alchemy for Monks and Friars," 245.

55 John Edwin Sandys stated that Bacon was imprisoned for almost 20 years on the orders of his superiors. However, the first document to report this imprisonment is the *Chronica XXIV Generalium Ordinis Minorum*, completed between 1369–1374. Bacon never mentions any imprisonment – the closest he comes is claiming that he has been unable to complete the *Opus maius* as quickly as he would have liked, because of the interference of his superiors. See Bacon, *Letter to Clement IV*, 143–174; Power, "Seeking Remedies for Great Danger," 63–78; and Sandys, "Roger Bacon."

56 Crowley, *Roger Bacon.*

57 Newman, "Roger Bacon and Secrecy."

58 Theisen, "The Attraction of Alchemy for Monks and Friars," 245.

Bibliography

Manuscripts

Roger Bacon. *Opus minus.* In Rome, Biblioteca Apostolica Vaticana MS Reginensis Latini 1317, folios 108v–132r.

Primary Sources

Roger Bacon. *Antidotarius.* Edited by Andrew G. Little and E. Withington. *Opera hactenus inedita Rogeri Baconi, Fasc. IX. De Retardatione Accidentium Senectutis cum aliis opusculis de rebus medicinalibus.* Oxford: Oxford University Press, 1928.

Roger Bacon. *De erroribus medicorum.* Edited by Andrew G. Little and E. Withington. *Opera hactenus inedita Rogeri Baconi, Fasc. IX. De Retardatione Accidentium Senectutis cum aliis opusculis de rebus medicinalibus.* Oxford: Oxford University Press, 1928.

Roger Bacon. "Epistola de secretis operibus artis et naturae et de nullitate magiae." In *Fr. Rogeri Bacon Opera Quaedam Hactenus Inedita, vol. I: Containing the Opus Tertium, Opus Minus, and Compendium Philosophiae*, edited by John S. Brewer. London: Longman, Green, Longman, and Roberts, 1859.

Roger Bacon. *Letter to Pope Clement IV.* Translated by Nikolaus Egel. *Revista Española de Filosofía Medieval*, 27/2 (2020): 143–174.

Roger Bacon. *Liber sex scientiarum.* Edited by Andrew G. Little and E. Withington. *Opera hactenus inedita Rogeri Baconi, Fasc. IX. De Retardatione Accidentium Senectutis cum aliis opusculis de rebus medicinalibus.* Oxford: Oxford University Press, 1928.

Roger Bacon. *The Opus Maius.* Translated by Robert Belle Burke. *2 vols.* Philadelphia: University of Pennsylvania Press, 1928.

Roger Bacon. *Opus Majus*. Edited by John Henry Bridges. *3 vols*. Oxford and Edinburgh: Clarendon, 1897–1900.

Roger Bacon. *Opus minus*. Edited by John S. Brewer. *Fr. Rogeri Bacon Opera Quaedam Hactenus Inedita, vol. I: Containing the Opus Tertium, Opus Minus, and Compendium Philosophiae*. London: Longman, Green, Longman, and Roberts, 1859.

Roger Bacon. *Opus tertium* Edited and Translated by Nikolaus Egel. Hamburg: Felix Meiner Verlag, 2020.

Roger Bacon. *Secretum secretorum*. Edited by Robert Steele. *Opera hactenus inedita Rogeri Baconi*, fasc 5. Oxford: Clarendon, 1920.

Secondary Sources

Allen, Meagan S. *Roger Bacon and the Incorruptible Human (1220–1292): Alchemy, Pharmacology, and the Desire to Prolong Life*. London: Palgrave Macmillan, 2023.

Crowley, Theodore. *Roger Bacon: The Problem of the Soul in His Philosophical Commentaries*. Louvain: Éditions de l'Institut supérieur de philosophie, 1950.

Gruman, Gerald J. "A History of Ideas about the Prolongation of Life: The Evolution of Prolongevity Hypotheses to 1800." *Transactions of the American Philosophical Society* 56: (1966): 1–102.

Hill, Carole. *Women and Religion in Late Medieval Norwich*. Woodbridge: Boydell and Brewer, 2010.

McVaugh, Michael. "The 'Humidum Radicale' in 13th Century Medicine." *Traditio* 30 (1974): 259–283.

McVaugh, Michael. "Theriac at Montpellier 1285–1325." *Sudhoffs Archiv, Zeitschrift fur Wissenschaftsgeschichte* 56/2 (1972): 113–144.

Newman, William R. "The Alchemy of Roger Bacon and the *Tres Epistolae* Attributed to Him." In *Comprendre et maitriser la nature au moyen age: melanges d'histoire offerts a Guy Beaujouan*, edited by Danielle Jacquart, 461–479. Geneva: 1994.

Paravicini Bagliani, Agostino. "Ruggero Bacone autore del *De retardatione accidentium senectutis*" *Studi Medievali* 28 (1987): 707–728.

Paravicini Bagliani, Agostino. "Ruggero Bacone e l'alchimia di lunga vita. Riflessioni sui testi." In *Alchimia e medicina nel Medioevo*, edited by Chiara Crisciani and Agostino Paravicini Bagliani, 37–54. Florence: SISMEL, 2003.

Power, Amanda. "Seeking Remedies for Great Danger: Contemporary Appraisals of Bacon's Expertise." In *Knowledge, Discipline and Power in the Middle Ages: Essays in Honour of David Luscombe*, edited by Joseph Canning, Edmund J. King, and Martial Staub, 63–78. Leiden: Brill, 2011.

Rasmussen, Seth C. *The Quest for Aqua Vitae: The History and Chemistry of Alcohol from Antiquity to the Middle Ages*. New York: Springer, 2014.

Sandys, Sir John Edwin. "Roger Bacon." *Proceedings of the British Academy* 6 (1913–1914): 371–388.

Tavormina, M. Teresa. "Roger Bacon: Two Extracts on the Prolongation of Life." In Sex, Aging, & Death in a Medieval Medical Compendium: Trinity College Cambridge MS R.14.52, its Texts, Language, and Scribe. Arizona Center for Medieval and Renaissance Studies, Tempe, AZ, 2006.

Theisen, Wilfrid. OSB. "The Attraction of Alchemy for Monks and Friars in the 13th–14th Centuries." *The American Benedictine Review* 46/3 (1995): 239–254.

Walker Bynum, Caroline. *Holy Feast and Holy Fast: The Religious Significance of Food to Medieval Women*. Berkeley: University of California Press, 1987.

Ziegler, Joseph. *Medicine and Religion c. 1300: The Case of Arnau of Vilanova*. Oxford: Clarendon, 1998.

9 Roger Bacon and Secrecy

Alchemical Techniques of Deception and the Doctor Mirabilis

William R. Newman

9.1 Techniques of Alchemical Secrecy

Roger Bacon's innovative position in the development of alchemical medicine is now widely acknowledged and has recently served as the basis of a groundbreaking book, but his role as an exegete, and even as a fashioner, of alchemical puzzles is less widely known.[1]

This is surprising because a close inspection of Bacon's sources shows that alchemy was fundamental to his view of secrecy in science, and it appears to have contributed in a major way to the very structure of his three great *Opera*. Furthermore, Bacon's quite self-conscious analysis and use of alchemical secrecy place him in a unique position. Unlike most other medieval aficionados of alchemical puzzles, Bacon not only made use of different techniques of secrecy but subjected them to a sort of second-order analysis in which he stood outside of the material and acted as a scholarly interpreter of the language of enigmata.

In addition to outright cryptography, alchemists traditionally employed multiple techniques of literary deception.[2] A number of these are now well-known, but little work has been done on the way these devices evolved over time. Let me mention five of them here, which I have found to occur in many alchemical works: the use of *Decknamen* or cover-names for alchemical ingredients; the intentional omission of operational stages, ingredients, or even parts of theories, which I call *syncope*; the opposite device, wherein these stages, ingredients, or theories are multiplied beyond necessity, and new names are given for the same thing; I call this *parathesis*. Additionally, alchemists employed language where the same terms may mean different things at different stages of the overall process of making the philosophers' stone. This might be considered a refinement of the *Decknamen* approach, but since it involves more than simple substitution of one name for another in a fixed process, I have named it "graduated iteration."[3]

Finally, there is *dispersa intentio*, which, unlike the first four techniques, is an actual actor's category first given the name *tabdīd al-ʿilm* (dispersion of knowledge) by Arabic authors.[4] This refers to the intentional splitting up and reordering of a discourse in order to confuse the reader. Although all of these techniques of deception have been identified in early modern alchemical

DOI: 10.4324/9781003405702-10

texts, little effort has been directed towards their appearance in the alchemy of the Middle Ages. How many of these devices were actually used in the medieval period, as opposed to being developments of a later time? The question is an important one if we are going to determine the course of alchemy's evolution over the *longue durée* and if we wish to avoid reading the older texts anachronistically. Because of his careful analysis and use of alchemical language in his three *Opera*, Roger Bacon provides an excellent case study for ascertaining the state of the art in the thirteenth century.

9.2 Bacon's's Justification of Secrecy

Let us begin with Bacon's best-known work, the *Opus maius,* and then pass to the other texts where he considers the topic of alchemical secrecy, and indeed, secrecy in general. Despite its wealth of information about the other sciences, the *Opus maius* is not a rich trove of alchemical knowledge. Alchemy appears here mainly in a handful of pages intended to show the need for subordinating the aurific art to the architectonic role played by *scientia experimentalis.* Alchemy can hardly attain the 24 carats of pure natural gold, for example, but when guided by experimental science, it can even make a gold of 30, 40, or as many degrees as we want.[5] Similarly, Bacon's oft-repeated story of a rustic who prolonged his life by drinking a liquid found in a golden bottle buried in a field leads him to say that this may have involved a gold better than that made by alchemy or nature. In order to maximise the medical effect of this gold by mixing it with other simples, *scientia experimentalis* is required.[6] Just as geometry, astronomy, and other sciences can do far more when directed by the *experimentator peritus,* so too with alchemy.

Bacon will say far more on alchemy in the *Opus minus* and *Opus tertium,* but there is still another, less obvious area where the aurific art touches the *Opus maius.* At the very beginning of this long book, where Bacon inveighs against the errors of the crowd (including the "rabble of philosophers," the *vulgus philosophantium*) in Part 1, Chapter 4, he provides an interesting description and justification of the secrecy employed by the wise. When Aulus Gellius supposedly advised that one should feed donkeys their rough thistle instead of treating them to lettuce, he was referring to the fact that the Greek *sapientes* did not reveal their wisdom to the masses.[7] Similarly, the Sermon on the Mount admonished that one should not spread pearls among swine (Matthew 7:6), and the *Liber de gemmis* by Marbode of Rennes enjoined that the revelation of secrets to the crowd (*turba*) would only diminish their majesty.[8] Applauding this advice, Bacon then passes to the words of the *Secretum secretorum* of pseudo-Aristotle. Explaining to his student Alexander the Great that he cannot reveal every single secret to him without reservation, the Aristotle of the *Secretum* says that such incautious openness would "make him a breaker of the celestial seal" (*se fore fractorem sigilli coelestis*). In order to avoid this sin, and to escape the possibility that the rabble would use these powerful *arcana* to turn the world topsy-turvy, the ancient wise men did not write their secrets "on the skins of goats and sheep" in a way that the vulgar could comprehend them. In fact, they sometimes reserved their highest secrets for oral communication, and even when they deigned to commit them to vellum, they often employed a variety of tools of deception.

In Bacon's view, it is not just the *Secretum secretorum* that conveys Aristotle's disdain for the *vulgus* and the consequent need for secrecy, but even the texts of the Organon. Thus, Aristotle's discussion of the probable status of conclusions reached by dialectic in the *Topics* is, for Bacon, an admonition against the *stultitia* of the crowd. Since Aristotle distinguishes there between the *plures* and the *sapientes* in Book I, Chapter 10 (= Bekker 104a) as part of this discussion, Bacon is provided with a pretext for launching into a page-long diatribe against the *vulgus*.[9] At best, the masses can accidentally happen upon truth, and when they do, their natural tendency is to pervert it. Hence, the best course of action is to reserve the secrets of the wise and keep them from the hands of the feebleminded multitude by either writing nothing at all or by employing "figurative writings and many other methods."[10]

The paradigmatic examples of these deceptive literary devices are the *enigmata* provided by the riddle of the philosophical egg that Aristotle relates in *pars tercia* of Bacon's *Secretum secretorum* edition, and by the *Tabula smaragdina* of Hermes Trismegistus (whom Bacon's text calls Hermogenes), both of which concern alchemy. It is worth reiterating the fact that it is alchemy rather than astrology, magic, or other secretive disciplines that supplies Bacon with his major source of verbal riddles. If one examines other examples of occult arts in Bacon's edition of the *Secretum secretorum*, the secrecy that one finds is in reality minimal. Consider the instructions for making the talisman of Hermogenes in Chapter 23.[11] The *Secretum* tells us there to take the substance of Saturn, Jupiter, Mercury, Sol, Venus, Mars, and Luna in equal parts, and to add additional Saturn and Sun. These planetary names are in fact alchemical *Decknamen* for the seven metals known at the time, lead, tin, quicksilver, gold, copper, iron, and silver. The talisman-maker is then supposed to fuse these together at a time when Leo is the ascendant, Sol is in its exaltation (Aries 19°), Luna in its exaltation (Taurus 3°), and the remaining five planets in one of their two respective houses, thus Saturn in Aquarius, Jupiter in Sagittarius, Mars in Scorpio (reading ♏ for ♍), Venus in Taurus, and Mercury in Virgo. None of this technical astrological information is encoded by the *Secretum secretorum*. The only element of secrecy here lies in the planetary *Decknamen* used for the metals. But Bacon himself clearly realises that these are properly alchemical terms, since he explains them in the chapter entitled "De expositione enigmatum alkimie" (Part 2, Chapter 113) of the *Opus tertium*. In short, the fact that the *Secretum secretorum* does not encode talismanic magic, but does so for alchemy, supports my view that the actual secrecy found in the *Secretum secretorum*, such as it is, belongs mostly to alchemy. Therefore, given the heavy influence that the

Secretum secretorum had on Bacon, we can say that the aurific art has tinctured his overall acroamatic view of the sciences in a quite significant way.

9.3 Dispersa Intentio

In fact, a significant influence from alchemy can even be detected in the odd way that Bacon used it in structuring his three *Opera*. In his new edition of the *Opus tertium*, Nikolaus Egel correctly points out that Bacon's motives for dividing his *Opera* into three separate books were twofold. On the one hand, the *Opus minus* and *Opus tertium* served to fill out parts of the *Opus maius* that Bacon felt he had left incomplete, and indeed the *Opus tertium* did the same thing for the *Opus minus*. Bacon effectively tells us this in the prefatory letter to Clement IV that begins the *Opus tertium*.[12] But as Egel points out, Bacon was also driven to divide up his works and pass them on piecemeal by the desire to keep secrets out of the hands of the vulgar.[13] This emerges with great clarity from Bacon's comments in Part 2, Chapter 112, which bear the title "De enigmatibus alkimie."

This very rich chapter first describes the need for secrecy in alchemy, repeating the adage that Bacon attributes to Aulus Gellius that one should not feed lettuce to donkeys. Bacon then asserts the well-worn *donum Dei* topos that God reveals the secrets of alchemy to His chosen by inspiration; hence they must not be vulgarised to the masses. This leads him to his reasons for dividing up the secrets of alchemy and presenting them to Clement *sparsim*. As he puts it:

> I have freely diversified so many writings for two reasons: the chief is that I might be able to open these marvels to your Holiness, just as is possible and appropriate at this time, and the second is that they be hidden from others. For these four writings will hardly fall into anyone's hands, and I do not care if anyone should see one, or the second or third, because unless he carefully applies himself to all four, he will be able to understand nothing of the very great secrets [therein].[14]

What Bacon is telling us is that he has written both to reveal and to conceal, a classic alchemical conceit.[15] But his particular method of attaining these contrasting goals is even more clearly derived from alchemical authors. In a word, Bacon is employing *dispersa intentio* or dispersion of knowledge, the *tabdīd al-ʿilm* of the Arabic alchemists, especially used by those writing under the *nom de guerre* of Jābir ibn Ḥayyān. This technique of deception consisted of dividing up one's discourse, be it a recipe or a discussion, into multiple parts and dispersing them, either in different parts of the same book or even in different books. Although employed by many alchemical writers, *dispersa intentio* was the hallmark of the Jabirian corpus, and it appeared in the *Liber de septuaginta* translated by Gerard of Cremona, which Bacon could well have read.[16] Jābir's Latin follower, the pseudo-Geber who wrote his influential *Summa perfectionis* in the last quarter of the thirteenth century, would famously rephrase the words of his master at the beginning and end of the *Summa*, saying:

> Whatever was hidden by us in one part we have made manifest in the corresponding part of this volume, so that the achievement of this both very excellent and noble part of philosophy be made accessible to the wise [...] let us relate that we have not passed on our science in a continuity of discourse, but that we have strewn it about in diverse chapters. This is because both the tested and the untested would have been able to take it up undeservedly, if the transmission were continuous.[17]

So we see pseudo-Geber advising precisely the very practice that Bacon himself employed when he divided his three *Opera* into three parts and sent a fourth, unnamed one to Clement by way of Young John. Although Bacon had probably not read the *Summa perfectionis* of the Latin pseudo-Geber, the fact that both authors imbibed from the same literary traditions provides evidence of the widespread diffusion of alchemical *dispersa intentio* in the thirteenth century. Other medieval texts of alchemy that Bacon definitely did read, such as the Latin *Liber de compositione alchimiae* of Morienus, also made use of *dispersa intentio* in practice, even if they lack the explicitly programmatic description of it given by Geber.[18]

Moreover, in the section of Brewer's defective, truncated edition of the *Opus minus* labelled "De rerum generationibus," which gives a comprehensive account of mineral generation, Bacon suggests that he has employed *dispersa intentio* not only when he divided the *Opera* into individual works, but even within the confines of the *Opus minus* taken by itself. After asserting that knowledge of mineral generation and of alchemy is necessary for the proper understanding of scripture, Bacon adds: "But also without these things the secrets of alchemy, which I hid above (*superius occultavi*), cannot be known." Bacon is likely referring here to the acephalous beginning of the *Opus minus* in Brewer's edition, which must be supplemented with a section that I published in 1995, based on the text of the *Opus minus* found in an important manuscript kept at the Vatican, *Vaticanus reginensis* 1317.[19] The entire section provides an extremely dense reading of the Philosophical Egg riddle found in the *Secretum secretorum,* which Bacon interprets as a *Deckname* for human blood. He has cobbled together a host of alchemical authorities including pseudo-Avicenna, pseudo-Rhazes, Morienus, and others to decode this riddle. But as the following passage, which occurs a few lines after Bacon's reference to his practice of *dispersa* intentio, suggests, the reader cannot really have a full understanding of the text without also acquiring the material on natural generation found near the end of the *Opus minus*. Once again, he is employing the technique of *dispersa intentio*:

> And it pleases me to treat these things in this place so that the secrets of alchemy may be hidden in these things that I write, but that your discretion may regard them. For unless the things that I wrote before be known, as well as those that I will write, and thirdly those that I will insert in the end of the work, the truth of these secret writings cannot be known.[20]

Having already shown Bacon's debt to alchemy for his overall approval of secrecy, we can now say that the Friar's approach to science as a literary form was heavily conditioned by his interaction with alchemical texts. Both his insistence on the need to conceal secrets and, more distinctly, his use of *dispersa intentio* to achieve that goal, permeate his three *Opera.* So far, however, my analysis of Bacon's work has focused on rather general characteristics of his literary output. If we now turn from his employment of literary techniques such as *dispersa intentio* to his more particular comments on the techniques of literary deception, a much richer picture emerges.

9.4 The Categorising of Techniques of Deception

Let us now examine Bacon's attempt to categorise different techniques of secrecy, some, but not all of them, derived from alchemy. The easiest way to achieve this goal is by looking at the *Epistola de secretis operibus artis et naturae,* edited rather haphazardly by Brewer in 1859. It is entirely possible that the first nine chapters of the *Epistola* as we have it were composed by a follower of Bacon rather than by Bacon himself, but if so, the follower for the most part merely cobbled together parts of the three *Opera* and possibly other Rogerian material.[21] And since Bacon himself was a serial self-plagiarist, this literary recycling in itself provides no evidence of inauthenticity. We need not comment for now on chapters ten and eleven, which are more complicated and raise their own issues.

Chapter 8 of the *Epistola* opens with themes that are by now familiar to us: the need for secrecy is forcefully presented, bolstered by Aulus Gellius's maxim about lettuce and donkeys, the admonitions of Marbode, and the *Secretum secretorum's* concerns about breaking the seal of heaven. The chapter even presents a fuller version of Bacon's peculiar exposition of Aristotle's *Topics* than the one we encountered in the *Opus maius.* In the *Epistola*, Aristotle's division of the multitude (the *plures*) from the wise (the *sapientes*) proceeds to a discussion of common knowledge (*communes conceptus*) versus the particular results arrived at by the arts and sciences starting from their own principles (*principia propria et conclusiones artium et scientiarum*). It is the latter that are of potential value, as they lead to the discovery of secrets rather than to mere commonplaces.[22]

Instead of focusing on this material, let me jump directly into the *Epistola's* analysis of scientific secrecy. Chapter 8 divides the literary techniques of concealment into the following seven categories, which we will briefly describe in the order in which the text presents them. According to the *Epistola*, the *sapientes* have ensured the restricted reception of their hidden arts in the following ways:

1. through the use of characters and verses.
2. by employing enigmatic and figurative words (e.g., the philosophers' egg).
3. by using different methods of writing, for example only consonants, "just as the Jews, Chaldeans, Syrians, and Arabs write their secrets."
4. through mixture of letters of different genus (like Ethicus).
5. by creating their own, personal alphabets (like Artephius).
6. through the use of geometrical figures, whose diverse "points and notes" signify different letters (like Artephius).
7. by using what seems to be an abbreviated version of scribal shorthand: the *ars notatoria*, "quæ est ars notandi et scribendi ea brevitate qua volumus, et ea velocitate qua desideramus."

The first category, where the *sapientes* hide their wisdom behind characters and verses, probably refers to Bacon's idea that natural secrets are buried within books of magic. A closely related notion appears in the *Opus maius* and *Opus tertium*, where Bacon deprecates the magicians' claim that their *carmina* and characters are responsible for natural effects like that of the lodestone to attract iron.[23] The *Opus maius* and *Opus tertium* also give the more colourful example of a hazel or willow shoot that is separated longitudinally and then spontaneously reunites its two halves because of the magician's *carmina*. By *carmina*, Bacon probably means incantations: this, after all, is one of the meanings that DuCange gives to the term. At any rate, the *Epistola* expands on this idea in its second chapter, saying that the genuine philosophers have intentionally hidden secrets of nature under magical verbiage in order to delude the unworthy. As in the *Opus tertium*, the *Epistola* uses the example of the lodestone, though the willow and hazel are absent there:

> Just as if it were wholly unknown that the magnet attracts iron, and someone wishing to carry this out before the public should make characters and bring forth incantations [*carmina*], lest it be perceived that the whole work of attraction is natural.[24]

This text of the *Epistola* expands in a subtle and interesting way on its parallel passages in the *Opus maius* and *Opus tertium*. Whereas the *Opus tertium*, in particular, presents the characters and *carmina* of magicians as evidence of their ignorance, since they are unaware that it is a natural property of certain materials to "run together" or concur (*concurrunt*), the *Epistola* suggests that magicians and perhaps others may simply be using the verbal and graphic apparatus of magic in an obfuscatory fashion. In other words, the author of the *Epistola* opens up the possibility that the *carmina* and characters of "magic" books are merely window-dressing that serves to obscure a deeper level of natural knowledge hidden within. It is this claim that gives him licence to include *characteres et carmina* among the techniques of deception outlined in Chapter 8 of the *Epistola*.

Let's now pass to the *Epistola's* second entry, "enigmatic and figurative words." While the *Epistola's* first literary deception appears to be drawn from magic books, the second relates directly to alchemy. Chapter 8's primary example of "enigmatic and figurative words" is precisely that of the philosophers' egg, the riddle that Aristotle relates to Alexander in the *pars tercia* of the *Secretum secretorum*. Here is what the *Epistola* says on the matter:

> Some have hidden many things by means of characters and incantations, but others by means of enigmatic and figurative words, as Aristotle says to Alexander in the *Book of Secrets*, saying to Alexander: "I wish to show you, Alexander, the greatest secret of secrets, and may the divine power aid you in hiding the arcanum, and at attaining the goal. Therefore take the stone that is not a stone, and which is in every man, and in every place, and in every time, and it is called the philosophers' egg, and the limit of the egg." And thus there are found innumerable things in many books and in diverse sciences obscured by such language, so that they may be understood in no way without a teacher.[25]

As I have shown elsewhere, the *Opus minus* combines pseudo-Avicenna's *De anima in arte alkimie*, the *Secretum secretorum*, and several other alchemical sources to arrive at the conclusion that the philosophers' egg refers to human blood, which must be divided into fractions and purified.[26] So, the second technique of obscuring scientific knowledge lies in the use of alchemical riddles.

A second riddle that Bacon analyses at length appears in his commentary on the Hermetic *Tabula smaragdina*, which is reproduced in full in the *Secretum secretorum*. For those who may be unfamiliar with this enigmatic text, I translate it quite literally as it appears in Bacon's Latin *Secretum secretorum*. I have followed Jean-Marc Mandosio's numeration of the stanzas in both the Latin and English translation for the sake of convenience:

> And our father Hermogenes [i.e. Hermes Trismegistus], who is triple in philosophy, said, in prophesying very well:
>
> 1. The truth is held <to be> thus, and there is no doubt that inferiors respond to superiors and superiors to inferiors. The worker of miracles [*Operator miraculorum*] is the one God alone, from whom every marvelous operation descends. Thus all things are generated from one sole substance, one sole disposition.

2. Whose father is Sol, whose mother is Luna. Which has borne nature itself in its womb, through the wind; the earth is impregnated by it.
3. Hence Sol is called the father of causes, the treasure of miracles [*thesaurus miraculorum*], the giver of virtues. Earth is made from fire. Separate the earthy from the fiery, because the subtle is more noble than the gross, and the rare [more noble] than the thick.
4. This is done wisely and discretely, for the earth ascends into heaven, and falls down from heaven to earth. And hence it destroys the upper and lower virtue.
5. So therefore it rules over superior and inferior [things], and you will rule over the upward and the downward, for the light of lights is with you, and on account of this all shadows will flee from you. For the upper virtue conquers all things. For every rare [thing] acts on every dense [thing].
6. And this operation proceeds according to the disposition of the Macrocosm.
7. And on account of this he is called Hermogenes triple in philosophy.

The precise meaning of the *Tabula smaragdina* is, of course, extremely hard to pin down, as most modern scholars who have worked on it agree.[27] What is important for us is Bacon's interpretation. The *lemmata* provided by the *Tabula* seem to have given him more food for thought than any other passages in the *Secretum secretorum,* including those dealing with Aristotle's marvellous medicine, the *Gloria inestimabilis.*[28] One might expect Bacon to interpret the *Tabula* simply as a veiled alchemical recipe, as his contemporary Albert the Great did.[29] But his interpretation is more subtle than that. To Bacon, the sun and moon in the *Tabula* can refer *both* to the celestial luminaries *and* to gold and silver. In glossing Stanza 2, he says that the Moon really does impregnate our planet by collecting the virtues of the stars, mixing them with her own, and then passing them on to the Earth. But that does not exclude an alchemical interpretation as well. Let us consider Bacon's interesting explanation here, from his commentary on the *Secretum secretorum*:

> This can be expounded alchemically and figuratively as [being] about gold and silver, for in human circumstances, all things come about through gold and silver. Hence "gold" can be understood by the word "sun," and "silver" by the word "moon"; and this is the custom among alchemists. And the "earth" [i.e. in Stanza 2] insofar as it relates to human affairs, is impregnated and fertilized by silver, and gold is the "Father of Miracles" etc., that is, by means of it, marvels are performed in human affairs.[30]

Thus Bacon, rather than extracting a specific recipe from the *Tabula smaragdina*, sees it as advocating the role of gold and silver in society. The same level of generality is maintained throughout his interpretation until he arrives at Stanza 5. Up to that point, he has interpreted terms like "ascent" and "descent" to refer to general alchemical operations such as distillation and sublimation. When he gets to the final sentence of Stanza 5, however, *Omne enim rarum agit in omne densum*, he takes this to refer to the ability of the alchemists' subtle metallic "medicine" to enter into base metals and convert them to gold or silver. So we are now in the realm of the alchemical elixir. But instead of trying to extract precise directions for making the elixir from the *Tabula*, Bacon now inserts an elaborate gloss on the Riddle of the Philosophers' Egg, as though it were simply part of the *Tabula.* It is as if Bacon believes Aristotle introduced the words of Hermogenes (i.e., Hermes) into the *Secretum secretorum* merely as a riddling support to the *further* riddle of the philosophical egg. For Bacon, it is the division and recombination of the egg that provides the path to practice, not the enigmatic words of the *Emerald Tablet.*

One could continue with Bacon's treatment of these alchemical riddles in his commentary on the *Secretum secretorum*, but we still need to consider the other techniques of secrecy described in the possibly spurious *Epistola.* The remaining techniques of secrecy, numbers three through seven, all involve the manipulation of individual letters in one form or another. Some of these techniques may qualify as actual cryptography in the modern sense, though the *Epistola* nowhere describes any shared key by which they may be deciphered.

The third technique of deception describes a method in which the wise men have hidden their knowledge by the simple device of omitting the vowels in words:

> They have hidden [their knowledge] in a third fashion, by methods of writing, namely through consonants alone, so that no-one could read them unless he knew the referents of the words [*significata dictionum*], just as the Hebrews, the Chaldeans, Syrians, and Arabs write their secrets; indeed, they write practically everything in this way for the most part, and thus there is great wisdom hidden among them, especially among the Hebrews, since Aristotle says in the forementioned book [the *Secretum secretorum*] that God gave them all wisdom before there were philosophers; and that all nations have received the basis of philosophy from the Hebrews.[31]

Apparently, the *sapientes* have copied the practice of omitting vowels from the Semitic languages. The *Epistola* adds that the Hebrews, Syrians, Arabs, and Chaldeans write virtually everything in this fashion. The author is, of course, aware of the fact that this non-vocalic feature is simply a characteristic of Semitic languages in general, but his implication seems to be that because the Hebrews in particular possessed so many secrets, it was necessary that they make use of this omission of vowels as an obfuscatory technique.[32] The reference to the *Secretum secretorum* does, in fact, echo comments that Bacon made in his commentary on that text: Aristotle himself imbibed his wisdom from the Old Testament and was educated by the prophets. Moreover, Bacon's commentary insists that the pagans learnt their astronomy, geometry, and arithmetic from the "holy and just Hebrews."[33] Similarly, both the *Epistola* and the *Secretum secretorum* commentary go on to list Albumasar (Abū Maᶜshar) and Josephus as authorities for the claim that philosophy and the sciences descended from the ancient Hebrews.[34]

The fourth technique consists of mixing "letters of different genus," as "Ethicus the astronomer" did when he used Hebrew, Greek, and Latin letters.[35] This is a *topos* that appears in the *Opus maius* and *Opus minus* as well, in almost identical words. Thus the *Opus maius* refers to "Ethicus the philosopher in his *Liber de divinis et humanis et naturalibus,* which he wrote in Hebrew, Greek, and Latin writing [*sermo*], on account of the greatness of the secrets."[36] Bacon is referring to the *Cosmographia* of Aethicus Ister, a well-known medieval forgery. The *Cosmographia* does, in fact, discuss this practice, and even provides Aethicus's own, manufactured alphabet, which modified Hebrew, Greek, and Latin letters to arrive at entirely new letter forms.[37]

The *Epistola's* fifth and sixth forms of deception are more problematic than the foregoing, because they rely on the obscure philosopher Artephius. There is a long trail of scholarship on this figure, who, according to Bacon, managed to live 1,025 years, but no manuscript has yet been shown unequivocally to be the source of this claim.[38] As Nicola Polloni has recently pointed out, two versions of a work attributed to Artephius, the *Clavis sapientiae maioris,* were printed in the multi-volume *Theatrum chemicum,* begun by Lazarus Zetzner, and this text is no doubt a translation from the Arabic. As Polloni also shows, the *Clavis* was known to a figure who significantly influenced Bacon, namely Robert Grosseteste, so it is possible that our Friar knew the text, even if it does not contain the macrobiotic claims that Bacon attributes to Artephius. At any rate, the *Epistola's* fifth tool of deception lies in the use of personal alphabets, which it attributes to Artephius's book *De secretis naturae.* As it happens, the version of the *Clavis* attributed to Alfonso X El Sabio does contain something like that, though to say more about it would require a careful examination of the manuscript traditions.[39]

The sixth technique for maintaining secrecy, which the *Epistola* also attributes to Artephius, consists of "geometrical figures which according to the diversity of points and notes have the power of letters" (*figuræ geometricæ, quæ secundum diversitatem punctorum et notarum habent literarum potestatem*). It is, of course, very difficult to know what the author has in mind, since the printed versions of the *Clavis sapientiae maioris* contain nothing of the sort. It is tempting to link the *Epistola's* comments to the medieval magical figures made of lines connected by small circles called "caractères dits 'à lunettes'" and perhaps even to the "celestial alphabets" made famous in the Renaissance, but without further information, there is little more that one can say.[40] Clearly the topic needs more study.

Finally, the *Epistola* gives a seventh technique of secrecy: "Seventh, there is a greater technique of obfuscating, which is given in the *ars notatoria*, which is the art of denoting and writing with whatever brevity we wish, and with whatever speed we wish."[41] Although the *Secretum* could be speaking here of some specialised form of tachygraphy invented specifically for purposes of deception, the passage could also simply be referring to a highly abbreviated scribal shorthand.[42] The term that the *Epistola* uses for this, *ars notatoria*, should not be confused with the art of learning new subjects by means of magical prayer, called *ars notoria*. The genuine Bacon is well aware of this pursuit, and he condemns it explicitly in the *Opus tertium*; the *Epistola* itself rejects the *ars notoria* in its third chapter.[43]

9.5 Conclusion

To summarise and come to a conclusion, then, it is clear that Roger Bacon knew a number of the traditional literary techniques of deception employed by alchemists of his time and earlier. The most striking of these is the *dispersa intentio* of the Islamic alchemists and their epigones. Bacon evidently used this technique both in dividing the three *Opera* themselves and in presenting material within the individual texts in the form of non-continuous discourse. In addition to his intimate acquaintance with *dispersa intentio*, he had a keen awareness of alchemical *Decknamen*, which he expounded at length in the *Opera* under the rubric of *aenigmata*. The alchemical riddles that he analysed consist of sequential *Decknamen* placed into the imperative or subjunctive form of a recipe. And if the *Epistola de secretis operibus* is a genuine work of Bacon's, we can say that he went well beyond mere exposition here too, for the concluding chapters of that work go so far as to use his idiosyncratic system of *Decknamen* to cast his knowledge into the form of recipes. As for parathesis and syncope, neither Bacon's acknowledged writings nor the *Epistola* explicitly refer to these devices, although the use of "characters and verses" that begins the list of techniques in Chapter 8 of the *Epistola* may be related. When the *Epistola* speaks of *sapientes* who hide their natural knowledge beneath a misleading verbiage of magical terms, this sounds like something akin to alchemical parathesis, though perhaps not identical to it. Unfortunately, I have not found many parallel passages in the *Opera* that develop this idea, so I leave it only as a possibility. At any rate, I hope to have shown that Bacon was a keen exegete of encoded alchemical writing, and that in some instances, he even incorporated its arcane tricks into his own writings. It would be a worthwhile project to carry this research forward and determine the degree to which alchemical techniques of secrecy may enter into Bacon's writings in the three *Opera* on optics, astronomy, mathematics, and natural philosophy as well.

Appendix

The Tabula smaragdina (Roger Bacon, *Opera hactenus inedita*, Robert Steele, ed., (Oxford: Clarendon Press, 1920), vol. 5, pp. 115–117)

Et pater noster Hermogenes qui triplex est in philosophia optime prophetando dixit:

1. Veritas ita se habet et non est dubium, quod inferiora superioribus et superiora inferioribus respondent. Operator miraculorum unus solus est Deus, a quo descendit omnis operacio mirabilis. Sic omnes res generantur ab una sola substancia una sua sola disposicione.
2. Quarum pater est Sol, quarum mater est Luna. Que portavit ipsam naturam per auram in utero, terra impregnata est ab ea.
3. Hinc dicitur Sol causatorum pater, thesaurus miraculorum, largitor virtutum. Ex igne facta est terra. Separa terrenum ab igneo, quia subtile dignius est grosso, et rarum spisso.
4. Hoc fit sapienter et discrete. Ascendit enim de terra in celum, et ruit de celo in terram. Et inde interficit superiorem et inferiorem virtutem.
5. Sic ergo dominatur inferioribus et superioribus et tu dominaberis sursum et deorsum, tecum enim est lux luminum, et propter hoc fugient a te omnes tenebre. Virtus superior vincit omnia. Omne enim rarum agit in omne densum.
6. Et secundum disposicionem majoris mundi currit hec operacio.
7. Et propter hoc vocatur Hermogenes triplex in philosophia.

Notes

1 For the most recent word on Bacon's medical work in relation to alchemy, see Allen, *Roger Bacon and the Incorruptible Human*. For Bacon's work in relation to his Arabo-Latin sources, see Moureau, "Elixir atque Fermentum," 277–325; one should also consult Moureau's impressive study and edition of pseudo-Avicenna's *De anima*: Moureau, *Le De anima alchimique du pseudo-Avicenne*.

2 The use of cryptography ranged from extremely simple letter substitutions in the High Middle Ages to complicated keyword ciphers in the seventeenth century. For an example of the former, see the *liber Hermetis de blchkmkb* in Steele and Waley Singer, "The Emerald Table," 485–501. For an example of the latter, see Principe, "Robert Boyle's Alchemical Secrecy," 63–74. For more on cryptography in magical and alchemical texts, see Láng, "Ciphers in Magic," 125–141.

3 Newman, *Newton the Alchemist*, 31–44.

4 Kraus, *Jābir ibn Ḥayyān*, vol. 1, pp. xxvii-xxxiii. I borrow the Latin term *dispersa intentio* from Heinrich Cornelius Agrippa von Nettesheim, who took the idea, if not the term, from the *Summa perfectionis* of Geber. See Perrone Compagni, "Dispersa Intentio," 160–177.

5 Bacon, *Opus maius*, vol. II, 214–215.

6 Bacon, *Opus maius*, vol. II, 210, 213. For more on *scientia experimentalis*, see the chapter by Jeremiah Hackett in this volume.

7 See Bacon, *Opus maius*, vol. I, 10. The editor of the Berlin edition of Aulus Gellius's *Noctes atticae*, Martin Hertz, could find nothing resembling Bacon's recounting of the asses-and-lettuce adage in the genuine *Attic Nights*. See Hertz, "Introduction to *A. Gellii noctium atticarum libri* xx," xxxviii–xxxix.

8 Marbode of Rennes, *Liber de gemmis*, 1738A. "Nam majestatem minuit qui mystica vulgat, Nec secreta manent, quorum fit conscia turba."

9 The Aristotelian *lemma* for Roger's discussion, as found in Boethius's translation of the *Topics*, runs as follows: "Est autem propositio dialectica interrogatio probabilis aut omnibus aut pluribus aut sapientibus, et his vel omnibus vel pluribus vel maxime notis, non extranea; ponet enim aliquis quod videtur sapientibus, nisi contrarium sit plurium opinionibus." This is *liber: 1, cap.: 10, pag.: 15, linea : 17* (Bekker: 104a), as found in the online *Aristoteles latinus* database at https://clt-brepolis-net.proxyiub.uits.iu.edu/ald/Browse/Works/Details?id=a2ee3c86-41ed-4637-bd5a-d9d226f49fd5&tocid=134&forperson=, accessed 10 March 2024.

10 Bacon, *Opus maius*, vol. I, 10. "Et propter hoc sapientes licet darent in scriptis radices arcanorum sapientiae, tamen ramos et fructus vulgo philosophantium non dederunt. Nam vel omiserunt scribere, vel per sermones figurativos et multis modis, de quibus non est ad praesens dicendum, occultaverunt."

11 Bacon, *Opera hactenus inedita,* vol. 5, 162–163. "Congrega substanciam Saturni et Jovis Mercurii et Solis et Veneris et Martis et Lune, et victoriam ejus qui dominatur in eis, et est tunc racio - sicut magister omnium, et pone de ipso auro secundum quantitatem (sive qualitatem) que est in operacione. Et ascendens sit Leo, et Sol sit in signo competenti potens et ascendens et illuminans in .19. gradu, et Homo in Aquario, et prepara Lunam et ymaginem (sive magistrum) ejus ut sit in Cancro [aliter, Tauro (*recte*)], et in tercio gradu ipsius, et Jupiter in Sagittario et Mercurius in Virgine et Venus in Tauro et Mars in Virgine et ipsa prosperata in decimo, et Mercurius in Virgine in honore suo. Et quod magis placet ei, ut adjungatur Luna Soli et Jovi in locis respicientibus (vel recipientibus). Si possibile est tibi, ut sit inclinacio ad Jovem, et illuminacio ad Hominem, et respiciant se ad invicem benigno aspectu, et quilibet eorum emittat radium suum super consortem suum post recepcionem luminis sui. Et cavendum est ne dominus medii celi sit remissus et relaxans cordas, et situs planetarum edificancium sit in gradibus sui ascensus."

12 Bacon, *Opus tertium*, 12.

13 Egel, "Introduction," XXXIV–XXXV.

14 Bacon, *Opus tertium*, 956. "Et gratis tot scripturas variavi propter duas causas. Principalis est ut Sanctitati Vestre possem aperire hec magnalia, sicut possibile est et decet ad hoc tempus. 2a est, ut ab aliis celentur. Nam vix cadent hec 4 scripta in manu alicujus; et non curo si unum, vel 2°, vel 3° videat quicunque; quia, nisi omnia 4 diligenter attendat, nichil poterit de maximis secretis intelligere. See also Bacon, *Opus tertium*, 86, where the theme of *dispersa intentio* also comes up: et has radices ego pono in Secundo Opere post intentionem Majoris Operis datam: sed et in hac tertia scriptura ponam exquisitius. Sed nec quod hic ponam potest intelligi sine aliis locis, nec illa sine eis quae hic pono; nec omnia haec dant intellectum, nisi sapientissimis et omnino perfectis in hac scientia; qui non sunt tres in hoc mundo. Deus enim occultavit semper a multitudine potestatem hujus scientiae: nam vulgus non solum nescit uti rebus dignissimis, sed etiam convertit in malum; nec etiam vulgus sapientum ad haec unquam potuit attingere, sed solum sapientissimi et omnino expertissimi. Nolui vero radices istarum duarum scientiarum ponere in Majore Opere, quia non proposui tunc scribere de eis. Sed postea in Minore Opere vidi opportunum esse, et scripsi quae videbantur mihi expedire."

15 For the complementary processes of alchemical revelation and concealment, see Newman, *Gehennical Fire,* 115–169.

16 Berthelot, "Liber de septuaginta translatus a magistro Renaldo Cremonensi de lapide animali," 310. "Fuit ergo necesse componere libros ex quibus exponam ea que promissa sunt ex meis verbis, et intentiones meas extraneas. et alienum quod occultavimus in locis diversis et scienciis aliis. Erunt enim hii summe librorum

nostrorum, qui promissi sunt et qui sequuntur. In hiis namque administravimus quidquid est necessarium."

17 Geber, *Summa perfectionis*, 633.

18 Bacon refers to the Morienus text in the fragment of the *Opus minus* that I published in 1995. See Newman, "The Philosophers' Egg," 100. For *dispersa intentio* in the Morienus text, compare pages 39 and 45 to one another in Stavenhagen, *A Testament of Alchemy*.

19 Newman, "Philosophers' Egg," pp. 99–101.

20 Bacon, *Opus minus*, 360. "Sed et sine his secreta alkymiæ, quæ superius occultavi, sciri non possunt. This is followed a few lines later by the following: Et placet mihi in hoc loco de his tractare uti secreta alkymiæ occultentur in his quæ scribo, quod tamen vestra discretio valeat contemplari; nisi enim sciantur quæ prius scripsi, et ea quæ hic scribam, et tertio ea quæ in fine operis conscribo, non potest horum sciri veritas scriptorum secretorum." See also the chapter by Meagan Allen in this volume.

21 This possibility was already suggested by Thorndike in 1915. See Thorndike, "Roger Bacon and Gunpowder," 799–800.

22 Bacon, *Epistola de secretis operibus artis et naturae*, 543.

23 Bacon, *Opus tertium*, 884–886.

24 Bacon, *Epistola de secretis operibus artis et naturae*, 525. "Sicut si omnino esset ignotum quod magnes traheret ferrum, et aliquis volens hoc opus perficere coram populo, faceret characteres et carmina proferret, ne perciperetur quod totum opus attractionis esset naturale."

25 Bacon, *Epistola de secretis operibus artis et naturae*, 544. "Nam aliqui per characteres et carmina occultaverunt multa, alii vero per verba ænigmatica et figurativa, ut Aristoteles in libro Secretorum dicens Alexandro: 'O Alexander, volo tibi ostendere secretorum secretum maximum, et divina potentia juvet te ad celandum arcanum, et ad faciendum propositum. Accipe igitur lapidem qui non est lapis, et 'est in quolibet homine, et in quolibet loco, et in quolibet tempore; et vocatur ovum philosophorum, et terminus ovi.' Et sic inveniuntur in libris multis et scientiis diversis innumerabilia, talibus sermonibus obscurata, ut nullo modo intelligantur sine doctore."

26 Newman, "Philosophers' Egg," 79–99.

27 See, for example, Ruska, *Tabula smaragdina*; and Mandosio, "La Tabula smaragdina e i suoi commentary medievali," 681–696. Mandosio provides Roger Bacon's version of the *Tabula smaragdina* with numbered stanzas on 692–693.

28 A rough word count reveals that Bacon devoted well over 700 words to his exposition of the *Tabula* on pages 115–118 of his edition of the *Secretum secretorum* (in Bacon, *OHI*, vol. 5), in contrast to the 300-plus words that he wrote on the *gloria inestimabilis* on pages 105–106. See Meagan Allen for the *Gloria inestimabilis*.

29 Albertus Magnus, *Book of Minerals*, p. 17.

30 Steele, *OHI*, vol. 5, p. 116. "Sic omnes res, scilicet, naturales, istud potest exponi alkimistice et figurative de auro et argento, quia in rebus humanis temporalibus omnia fiunt per aurum et argentum. Unde per solem potest intelligi aurum et per lunam argentum, et hic est mos alkimistarum. Et terra quantum ad res humanas impregnatur et fecundatur per argentum, et aurum est pater miraculorum, et cetera, id est, per ipsum fiunt mirabilia in humanis rebus."

31 Bacon, *Epistola de secretis operibus artis et naturae*, 544. "Tertio modo occultaverunt per modos scribendi, scilicet per consonantes tantum, ut nemo posset legere, nisi sciat significata dictionum, sicut Hebræi, et Chaldæi, et Syri, et Arabes scribunt secreta; immo quasi omnia pro majori parte sic scribunt; et ideo est magna sapientia occultata apud eos, et maxime apud Hebræos; quoniam Aristoteles dicit

libro memorato, quod Deus dederit eis omnem sapientiam antequam fuerunt philosophi; et ab Hebræis omnes nationes habuerunt philosophiæ principium."

32 On this point, see Benoît, "Systèmes d'écriture, sémiotique et langage chez Roger Bacon," 97–98.

33 Bacon, *Secretum secretorum*, 62, n. 5. "Sciendum ergo quod infideles philosophi non invenerunt hanc scienciam nec alias partes mathematice, ut Geometriam et Arithmeticam, nec alias sciencias, set Deus dedit eas suis sanctis et justis Hebreis, a quibus omnes philosophi infideles habuerunt omnium scienciarum principia, ut dicit capitulo primo secunde partis hujus libri."

34 Bacon, *Epistola de secretis operibus artis et naturae*, 544. "Et Albumasar in libro Introductorio Majori, et alii philosophi, et Josephus primo et octavo Antiquitatum libris hæc manifeste docent." The more detailed parallel passage may be found in Bacon, *Secretum secretorum*, 62–63, n. 5.

35 Bacon, *Epistola de secretis operibus artis et naturae*, 544–545. "Quarto accidit occultatio per mixtionem literarum diversi generis. Nam sic Ethicus astronomus suam occultavit sapientiam, eo quod literis Hebræis, et Græcis, et Latinis eam conscripsit in eadam serie scripturæ."

36 Bacon, *Opus maius*, vol. II, 231. "Et Ethicus philosophus in libro suo de divinis et humanis et naturalibus, quem Hebraeo sermone, Graeco, et Latino, propter secretorum magnitudinem conscripsit, ponit in Deo Patrem, et Verbum Patris, et Spiritum Sanctum; et quod sunt tres Personae, Pater scilicet et Filius et Spiritus Sanctus."

37 Herron, *The Cosmography of Aethicus Ister,* 127, line 5; p. 147, l.14; especially pp. 163–165, ll. 15–20, and Aethcius's own alphabet on pp. 214–217. See also Grévin and Véronèse, "Les 'caractères' magiques au Moyen Âge," 373–374.

38 For the present state of research on Artephius, see Moureau, "Elixir atque Fermentum," 324–325.

39 Polloni, "A Matter of Philosophers and Spheres," 137–139. See Zetzner (ed). *Theatrum chemicum*, vol. 5, p. 877. Polloni refers to the text attributed to Alfonso El Sabio as the "AR-Version." A useful collection of medieval and early modern authors citing Artephius may be found in Carusi, "Il trattato di filosofia alchemica Miftah al-hikma ed i suoi testimoni presso la Biblioteca Apostolica," 66–77.

40 For "caractères dits 'à lunettes'," see Grévin and Véronèse, "Les 'caractères' magiques au Moyen Âge," 340, 341, 367, and 372. See also Le Pape, *Les écritures magiques,* 71–88. For the much later *scriptura caelestis* of Cornelius Agrippa, see his *De occulta philosophia,* 273–274.

41 Bacon, *Epistola de secretis operibus artis et naturae*, 545. "Septimo est majus artificium occultandi, quod datur in arte notatoria, quæ est ars notandi et scribendi ea brevitate qua volumus, et ea velocitate qua desideramus."

42 It is worth pointing out that the adjective *notatorius* is extremely rare. The only example of it in the sense used by Bacon found in the online Brepols *Database of Latin Dictionaries* is from the very passage of the *Epistola* cited here. See the Brepols *Database of Latin Dictionaries* at https://clt-brepolis-net.proxyiub.uits.iu.edu/dld/Dictionaries/Search?field=HEAD&query=notatorius, consulted 13 March 2024.

43 Bacon, *Opus tertium,* 882; Bacon, *Epistola de secretis operibus artis et naturae*, 545.

Bibliography

Primary Sources

Aethicus Ister. *The Cosmography of Aethicus Ister: Edition, Translation, and Commentary*. Edited by Michael Herron. Turnhout: Brepols, 2011.

Albert the Great. *Book of Minerals*. Translated by Dorothy Wyckoff. Oxford: Clarendon, 1967.

Aulus Gellius. *A Gellii noctium atticarum libri XX*. Edited by Martin Hertz. Berlin: Wilhelm Hertz, 1885.

Boethius. *Topics*. Accessed via *Aristoteles latinus database*. https://clt-brepolis-net.proxyiub.uits.iu.edu/ald/Browse/Works/Details?id=a2ee3c86-41ed-4637-bd5a-d9d226f49fd5&tocid=134&forperson=*Brepols Database of Latin Dictionaries Online* at https://www.brepols.net/series/dld-o

Cornelius Agrippa. *De occulta philosophia*. Cologne, 1533.

Geber. *Summa perfectionis*. Edited by William R. Newman. Leiden: Brill, 1991.

Lazarus Zetzner, ed. *Theatrum chemicum, vol. V*. Strassbourg: Zetzner, 1622.

Marbode of Rennes, *Liber de gemmis, Migne, Patrologia Latina*. Paris: Jean J. Bourassé, 1854.

Roger Bacon. "Epistola de secretis operibus artis et naturae et de nullitate magiae." In *Fr. Rogeri Bacon Opera Quaedam Hactenus Inedita, vol. I: Containing the Opus Tertium, Opus Minus, and Compendium Philosophiae*, edited by John S. Brewer. London: Longman, Green, Longman, and Roberts, 1859.

Roger Bacon. *Opus majus*. Edited by John Henry Bridges. *3 vols*. Oxford and Edinburgh: Clarendon, 1897–1900.

Roger Bacon. *Opus minus*. Edited by John S. Brewer. *Fr. Rogeri Bacon Opera Quaedam Hactenus Inedita, vol. I: Containing the Opus Tertium, Opus Minus, and Compendium Philosophiae*. London: Longman, Green, Longman, and Roberts, 1859.

Roger Bacon. *Opus tertium*. Edited and Translated by Nikolaus Egel. Hamburg: Felix Meiner Verlag, 2020.

Roger Bacon. *Opus tertium*. Edited by John S. Brewer. *Fr. Rogeri Bacon Opera Quaedam Hactenus Inedita, vol. I: Containing the Opus Tertium, Opus Minus, and Compendium Philosophiae*. London: Longman, Green, Longman, and Roberts, 1859.

Roger Bacon. *Secretum secretorum*. Edited by Robert Steele. *Opera hactenus inedita Rogeri Baconi*, fasc 5. Oxford: Clarendon, 1920.

Secondary Sources

Allen, Meagan S. *Roger Bacon and the Incorruptible Human (1220–1292): Alchemy, Pharmacology, and the Desire to Prolong Life*. London: Palgrave Macmillan, 2023.

Berthelot, Marcellin. "Liber de septuaginta translatus a magistro Renaldo Cremonensi de lapide animali." In *Archéologie et histoire des sciences*. Paris: Gauthier-Villars, 1906.

Carusi, Paola. "Il trattato di filosofia alchemica Miftah al-hikma ed i suoi testimoni presso la Biblioteca Apostolica." *Miscellanea Bibliothecae Apostolicae Vaticanae* 9 (2002): 35–84.

Grévin, Benoît and Véronèse, Julien. "Les 'caractères' magiques au Moyen Âge (XIIe-XIVe siècle)." *Bibliothèque de l'école des chartes* 162 (2004): 305–379.

Grévin, Benoît. "Systèmes d'écriture, sémiotique et langage chez Roger Bacon." *Histoire Épistémologie Langage* 24 (2002): 75–111.

Kraus, Paul, *Jābir ibn Ḥayyān: Contribution a l'histoire des idées scientifiques dans l'Islam. Mémoires présentés à l'Institute de'Égypte, 45, vol. 1*. Cairo: Imprimerie de l'institut francais d'archéologie orientale, 1943.

Láng, Benedek. "Ciphers in Magic: Techniques of Revelation and Concealment." *Magic, Ritual, and Witchcraft* 10/2 (2015): 125–141.

Le Pape, Gilles. *Les écritures magiques*. Milano: Arché, 2006.

Mandosio, Jean-Marc. "La Tabula smaragdina e i suoi commentary medievali." In *Hermetism from Late Antiquity to Humanism,* edited by Paolo Lucentini, Ilaria Parri, and Vittoria Perrone Compagni, 681–696. Turnhout: Brepols, 2003.

Moureau, Sebastien. *Le* De anima *alchimique du pseudo-Avicenne.* Florence: SISMEL, Edizioni del Galluzzo, 2016

Moureau, Sebastien. "Elixir Atque Fermentum: New Investigations about The Link between Pseudo-Avicenna's Alchemical De Anima and Roger Bacon: Alchemical and Medical Doctrines." *Traditio* 68 (2013): 277–325.

Newman, William R. *Gehennical Fire.* Cambridge, MA: Harvard University Press, 1994.

Newman, William R. *Newton the Alchemist.* Princeton: Princeton University, 2019.

Newman, William R. "The Philosophers' Egg: Theory and Practice in the Alchemy of Roger Bacon." *Micrologus* 3 (1995): 75–101.

Perrone Compagni, Vittoria. "'Dispersa Intentio.' Alchemy, Magic and Scepticism in Agrippa." *Early Science and Medicine* 5 (2000): 160–177.

Polloni, Nicola, "A Matter of Philosophers and Spheres: Medieval Glosses on Artephius's Key of Wisdom." *Ambix* 67 (2020): 135–152.

Principe, Lawrence M. "Robert Boyle's Alchemical Secrecy: Codes, Ciphers, and Concealment." *Ambix* 39 (1992): 63–74.

Ruska, Julius. *Tabula smaragdina: Ein Beitrag zur Geschichte der hermetischen Literatur.* Heidelberg: Carl Winter's Universitätsbuchhandlung, 1926.

Stavenhagen, Lee. *A Testament of Alchemy.* Hanover, New Hampshire: Brandeis University Press, 1974.

Steele, Robert and Waley Singer, Dorothea. "The Emerald Table." *Proceedings of the Royal Society of Medicine* 21/3 (1928): 485–501.

Thorndike, Lynn. "Roger Bacon and Gunpowder." *Science* 42/1092 (1915): 799–800.

Index

acceleration, centre of revolution 64
accidental modes of philosophy 43–44; and mathematics 47–49
addressativity 136
Africanus, Constantinus 97
agens 119
aging 147–148; *see also prolongatio vitae*
agriculture 74
air 151
Albert the Great 115–123, 135, 138, 173
Alberti, Leon Battista 106
alchemical language 166
alchemical riddles 172–173
alchemical secrecy 8–9, 165–166; *dispersa intentio* 168–170; justification for 166–168; techniques of deception 170–176
alchemy 8–9, 30–31, 74, 147, 150 154, 157–158; *see also* alchemical secrecy
Alexander the Great 166
al-Fārābī 47–48
al-Fārisī, Kamal al-Dīn 97
Alhazen 55–56, 76, 94; celestial influences 122–123; intentional properties of objects of vision 102–103; *Optics* 106n4; *perspectiva artificialis* 104–106; *perspectiva naturalis* 105; philosophy of mind 100, 103; physiology and neurology of vision 94–99
alhorod 47–48
Almagest (Ptolemy) 13, 15–16, 31–32, 44
animals: external sensorial stimulus 99–100; medicine 149–150
Antichrist 3
Antidotarius (Bacon) 150, 152
Aquinas, Thomas 44–45, 138
Archimedes, statics 57
Aristotelian framework 131
Aristotle 14, 22, 76, 111, 166–167, 170; *De caelo* 61; *De metheora* III (Aristotle) 27–28; deception 174–175; demonstration 77; mathematics 16; *Metaphysics* 43, 45–47, 76, 148; *Meteorologies* 28–29, 32; philosophy 43; *Posterior Analytics* 22, 61, 74–75, 77, 131
arithmetic 6–7, 72–73, 86–88; applicability of 78–79; division of 81–82; Pecham's view of 84–86; primacy of geometry 82–84; speculative arithmetic 81–82; *see also* mathematics
Arnulf of Provence 47
art, secrets 131–132, 138
Assub 14
astrolabe 19, 27, 29–30
astrologia practica 110–111, 119
astrologia speculativa 110, 117–118
astrological interpretations or judgements, defence of 112–115
astrological seals 135
astrology 31, 74, 109; defence of 112–115; experimental astrology 15; iudiciaria-operativa distinction 111–112; judicial astrology 15; and magic 111–112; and mathematics 109–111
astronomia iudiciaria 7, 111
astronomia operativa 7, 111
astronomy 7, 30–32, 79, 111; planets 167; *see also* astrology
atomism 73, 82, 84, 86
auctoritates 75
Avicenna 15, 28, 46; *De anima* 97

Bacon's framework 131, 136
Bacon's oeuvre 150–154, 158
bad secrecy 130, 132

bile 151
blood 151
Boethius 45, 59, 75, 81–82; *De arithmetica* 86
Book of Causes 27
Book of Sufficiency (Ptolemy) 15
burning mirror 19

camera obscura 97
Campanus de Novara 72
carmina 171
catoptromancy 134
celestial bodies 109–110
celestial influences 119, 122
Centiloquium (Ptolemy) 14, 16, 134
centre of revolution 61–65
certification, experimental science 18
characters 135
charms 135
Christendom 73
Christian theology 73
classes of senses 99
classification of the sciences, mathematics 73–75
Clavis 175
Clement IV 1–5, 149
cognition 103; faculties of 100
comets 14
Commentarii (*Commentaries*) (Ghiberti) 106
common sense 7, 96, 100–101, 102; visual perception 99
Communia mathematica (Bacon) 15–16, 31, 56, 59, 77, 79–81, 134
Communia naturalium 62, 74, 79–80
compound humors 151
concealment 138; *see also* secrecy
conjunctiva (*multahima*) 98
consolidativa 98
constellations 133–135
continuous quantities 83
corena 98–99
Corpora adae et evae (Bacon) 156
corpus equale 147–148, 152, 155–157
Cosmographia (Aethicus Ister) 175
cover-names for alchemical ingredients 165
crystalline humour 96–97, 100

d'Aguilon, François 95, 106
De anima (Avicenna) 97
De arithmetica (Boethius) 86
De caelo (Aristotle) 61
De canonio (*On the Balance*) 57
De erroribus medicorum (Bacon) 150
De Febribus (Isaac) 27
De imaginibus (Thebit) 134
De metheora III (Aristotle) 27–28
De mineralibus (Albert the Great) 135
De natura loci 118, 121–123
De numeris misticis (Pecham) 85–86
De ortu scientiarum (Kilwarbdby) 42
De ponderibus 59, 64
De ratione 58–59, 61, 63–64
De regimine senum 155
De retardatione accidentium senectutis 149–150, 155–156
De scientia experimentali (Bacon) 28–32; Chapters I–III 20–28
De scientiis (al-Fārābī) 16, 47–49
de secretis operibus artis et naturae et de nullitate magiae 26
De sphaera 79
De triangulis (Jordan) 67
death 148
deception, alchemical secrecy 170–176
Decknamen (cover-names for alchemical ingredients) 8, 165, 167
Della Pittura (*De Pictura*; *On Painting*) (Alberti) 105
demons 26, 132
demonstration 77
Desargues, Girard 106
diagrams 55, 60; centre of revolution 61–65; *see also* optics
diamonds, false empirical claims 23
diffusion of light 84
direct vision 94–97, 100, 104
discernment 104
dispersa intentio 9, 165, 168–170
divine science 44–45
division of arithmetic 81–82
division of mathematics, in *Communia mathematica* 79–81
donum Dei topos, alchemical secrecy 168
dual logic 76–77

earth 151
eclipse 14
Elementa (Jordan) 58
Epistola de magnete (Peregrinus) 136
Epistola de secretis operibus (Bacon) 26–27, 174
Epistola de secretis operibus artis et naturae (Bacon) 170–176

Epistola de secretis operibus et de nullitate magiae (Bacon) 150
equilateral triangles 21
essential modes of philosophy 43–44; and mathematics 44–47
estimation 99–102
Euclid 59, 79, 81–83
experience 6, 12, 20–24
experiential knowledge 24
experiential verification 22
experiment 136–138
experimental astrology 15
experimental science 5–6, 8, 12–16, 25–26, 73–74, 130, 133; alchemy 153; burning mirror 19; certification 18; *De scientia experimentali* 20–28; limitations of 16–17; magic 133; secrets 132
experimentation, optics 97
Experimentator Fidelis 17–20
external sensorial stimulus 99–100
eyes *see* vision

faculties of cognition 100
false empirical claims 23
false mathematics 113–115
fascination 132–133
fear 99
fidelis experimentator 17–20
fire 151
flowchart of mathematical knowledge 80
Foulques *see* Clement IV
founder of science 12–13
four humors 151
frameworks: Aristotelian framework 131; Bacon's framework 131, 136; scholastic framework 131
Franciscan order 2, 72
Franciscanism 130
freezing water 23

gaining knowledge 6, 20–21
Galle, Theodoor 95
Gellius, Aulus 166, 170
general science 79
geography 117
geomancy 13
geometry 21–22, 55, 72, 79, 87; burning mirror 19; equilateral triangles 21; motion 59; primacy over arithmetic 82–84; *see also* arithmetic; mathematics

Gerard of Cremona in Toledo 47–48
Ghiberti, Lorenzo 106
glacialis 98–99
Gloria inestimabilis 173
gold 152–153, 155, 173; medical effect of 166
good secrecy 130
graduated iteration 9, 165
grammar 47–48
gravity *see* positional gravity
Grosseteste, Robert 72, 83, 85
Gundissalinus, Dominicus 47, 56

halos 16–17, 28–29
heavenly bodies 31
Hermogenes 167, 172, 177
humors 151
humour albugineus 98–99
humour *vitreus* 98–99
hydromancy 13

illuminative knowledge 24
imagination 100–103
incantations 171
indivisiblism 82–83
intentional properties of objects of vision 102–103
interior knowledge 25
internal illumination 6
internal knowledge 24
isomorphism 97
iudicia astronomiae 112
iudiciaria-operativa distinction, astrology 111–112

John of Tynemouth 72
Jordan of Nemore 55–58, 81–82; application of his methodology 65–68; science of weights 56–59
judicial astrology 15
justification for secrecy, alchemy 166–168

Kilwardby, Robert 42
knowability 136
knowledge, gaining 6, 20–21

language, poetical language 48–49
Laon, Raymond of 4
lemmata 173
Liber de compositione alchimiae (Morienus) 169
Liber de gemmis (Marbode of Rennes) 166

Liber de septuaginta (Gerard of Cremona) 9, 168
Liber de triangulis Iordani (Jordan) 67
Liber karastonis (Qurra) 57
Liber pantegni (Constantinus Africanus) 97
Liber sex scientiarum (Bacon) 150, 157
life of Roger Bacon 1–2
light 95–96; diffusion of 84; reflection 101; refraction 101
limitations of experimental sciences 16–17
lines 82–83
literary deceptions, alchemical secrecy 165
logic 48, 60
logical arguments 6, 20–28
logical reasoning, mathematics as 75–78

magic 8, 13, 112, 114, 130, 132–136; and astrology 111–112; natural magic 132
magical astrology 115
magical blooks 26–27
magicians 26, 138
Magister Petrus 17–20, 27
Magister Speculi 134
magnets 27, 136; occult property 131; self-moving magnetic sphere 136–137
Marbode of Rennes 166
Marsh, Adam 72
Master Nicolas 72
Master Peter Peregrinus of Maricourt *see Magister Petrus*
mathematical method 60
mathematics 15–16, 21–22, 42–43, 75; and accidental modes of philosophy 47–49; and astrology 109–111; classification of the sciences 73–75; *Communia mathematica* 79–81; and essential modes of philosophy 44–47; as logical reasoning 75–78; and politics 45; speculative arithmetic 59; true versus false 112–115; *see also* arithmetic; diagrams; geometry
measurement systems 83
Mechanical Problems (Archimedes) 57
medicine 74, 149–150; humors 151; secrecy 150; secret seven 154–157; snake flesh 152; *see also* alchemy
melancholy 151
memory 99–102
Metaphysics (Aristotle) 22, 42, 43, 45–47, 76, 78, 85, 148
Meteorologies (Aristotle) 28–29, 32
mirrors: burning mirror 19; catoptromancy 134
mixed motion 59
modes of knowing 43
Montfort, Simon de 4
moral philosophy; *see also* Christian theology
moral teachings 24–25
motion 59, 66, 84–85
multiplication of species 118–119, 133
music 48, 79

Natural Astronomy 13
natural magic 132
natural occult 138
natural philosophers 29
natural philosophy 74
nature of place 121–123
nature, secrets 131–132, 138
neurology of vision 94–99
non-addressativity 135
numbers 85–86
numerus numerans 86
numerus numeratus 86

occult arts 167
occult property 131, 138
occultation 138–139; *see also* magic
occultation in magic 132, 138
Opera (Bacon) 5, 8, 150; alchemical language 166; *dispersa intentio* 168–170; secret seven 154–157
opera sapientiae 115
optic nerves 99
optical nervous system 100–101
optical system 94–99
Opticorum libri sex (d'Aguilon) 95
Optics (Alhazen) 7, 55, 67, 84, 94–99, 106n4; *see also* diagrams
Opus maius 2–3, 5; alchemy 152–154; astrological interpretations or judgements 112; astrology 112–115; Bacon's oeuvre 150–154; *De scientia experimentali* 20–32; justification for secrecy 166–168; magic 132–134; personal limitations 16–17; *perspectiva* 117; pharmacology 154–157; *prolongatio vitae* 147–150; secrecy 171; self-moving magnetic sphere 137

Opus minus 5, 31, 153–154; alchemical secrecy 156–158, 168–170; alchemy 147, 154, 166; *dispersa intentio* 168–170; experimental science 16–17; secrecy 172
Opus tertium 2, 8, 49, 73–74; alchemy 153–154; demons 26; *dispersa intentio* 168–170; experimental science 134–135; *Magister Petrus* 19; mathematics 109; medicine 154; *scientia experimentalis* 17; secrecy 167–168, 171; Unnamed Master 115–120

parathesis 9, 165
patiens 119
Pecham, John 72–73; view on arithmetic 84–86
Peregrinus, Peter (Pierre Pelerin de Maricourt) 136–137
perspectiva 12–13, 23, 28–29, 55, 73–74, 78, 84, 94–95, 117–119; common sense 96; experimental science 16–18; Unnamed Master 116
perspectiva artificialis 104–106
perspectiva naturalis 105
persuasio 5
Peter of Maricourt 72
phantasia 102
pharmaceuticals 8, 155–156
pharmacology 8, 147, 154–157
phenomenology of vision 103–104
philosophers 138–139
Philosophia moralis (Bacon) 46
Philosophical Egg riddle 169, 174
philosophical images 135–136
philosophy 6; accidental modes of 43–44; essential modes of 43–47; practical philosophy 45–47, 49; *see also* moral philosophy
philosophy of mind, Alhazen 100, 103
phlegm 151
Physics 42, 121
physiology of vision 94–99
place 121–123
planetary patterns of influence 119–120
planets 167
Platonism 76
poetical language 48–49
pointillism 96
points 82–83
politics, and mathematics 45
positional gravity 58–59, 65
Posterior Analytics (Aristotle) 22, 61, 74–75, 77, 131
power of reason 22–23
powers of sciences 3
practical astrology 110–111, 119
practical astronomy 31–32
practical geometry 81
practical philosophy 45–47, 49
prestidigitation 132
prolongatio vitae (prolonging of life) 8, 147–150, 153, 156–157
propter quid 59–60
pseudo-Aristotle 166–169
pseudo-Geber 168–169
psychology of vision 99–102
Ptolemy 13–14, 30; *Almagest* 13, 15–16, 31–32, 44; *Book of Sufficiency* 15; *Centiloquium* 16; mathematics 44; *Quadripartitum* 15
pyramids of force 119
pyromancy 13

Quadripartitum (Ptolemy) 6, 15
quadrivium (arithmetic, geometry, astronomy, and music) 72–73
quantities 60, 79; continuous quantities 83
Qurra, Thabit ibn 57

R01.2 61–62
rainbow 12, 16–17, 28–29, 97
reason 22–23, 47
recognition 103–104
reflection 101
refraction 101
resurrection 156
rhetoric 49
RI.003 65
RI.004 65
RI.006 65
Rubens, Peter 95

sapientes 166–167; secrecy 171; techniques of deception 170–176
sapientes mundi 85
scholastic framework 131
Scholastic method of commenting 60
science of weights 56, 74; Bacon, Roger 59–66; Jordan of Nemore 56–59
scientia experimentalis see experimental science
scientific utopia 136, 139–140
Scot, Michael 43

scriptum principale 5
secondary stars 14
secrecy/secrets 8, 130–131, 138; magic 132–136; *see also* alchemical secrecy
secret seven 150; disappearance of 157–158; *Opera* (Bacon) 154–157
Secretum secretorum (Bacon) 13–14, 18, 26, 150, 166–167, 169, 172; astrology 111; secrets 131–132; true mathematics 114
self-moving magnetic sphere 136–137
Seneca 28
sense-knowledge 24
senses 7, 99; external sensorial stimulus 99–100
simple humors 151
simple motion 59
snake flesh 152
Socrates 75
solids 82
speculative arithmetic 59, 81–82
speculative astrology 31
Speculum astronomiae 134
statics 56–57
stellar comets 14
stones, as medicine 152
straight lines 82
substantial method 77
Summa perfectionis 169
Summulae dialectices (Bacon) 46
surfaces 82
syncope 8–9, 165

tabdīd al-cilm 165, 168
Tabula smaragdina (Hermes Trismegistus) 167, 172–173, 176–177
talisman 7, 111–112, 134–135, 167–168
Tanqih al-manāzir (Recension of the Optics of Alhazen) 97
techniques of deception, alchemical secrecy 170–176
theologians, mathematics 113–115
Tractatus de experientia (Bacon) 21–26
Tractatus de experientia in communi (Bacon) 28–32
Trismegistus, Hermes 167
true mathematics 15, 111, 113–115
trustworthy experimenter 17–20
Tyrians 155

universal method 77
Unnamed Master 115–120
uvea 98

Vaticanus reginensis 169
velocity 57, 61
virtual velocities 57
virtus phantastica 102
vision: direct vision 94–97, 100, 104; intentional properties of objects of 102–103; *perspectiva artificialis* 104–106; phenomenology of 103–104; physiology and neurology of 94–99; psychology of 99–102
visual perception 7, 103; common sense 99
vitreus humour 98
vulgus 167

water 23, 151

Kolibri 360 GmbH
Ferdinand-Jühlke-Straße 7
99095 Erfurt, Deutschland
produktsicherheit@kolibri360.de